Nutrition and the Eye

Developments in Ophthalmology

Vol. 38

Series Editor

W. Behrens-Baumann   *Magdeburg*

KARGER

Nutrition and the Eye

Basic and Clinical Research

Volume Editor

Albert J. Augustin Karlsruhe

20 figures, 5 in color, and 17 tables, 2005

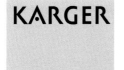

Basel · Freiburg · Paris · London · New York ·
Bangalore · Bangkok · Singapore · Tokyo · Sydney

....
Albert J. Augustin
Augenklinik
Moltkestrasse 90
DE-76133 Karlsruhe (Germany)

Library of Congress Cataloging-in-Publication Data

Nutrition and the eye / volume editor, Albert J. Augustin.
       p. ; cm. – (Developments in ophthalmology, ISSN 0250-3751 ; v. 38)
    Includes bibliographical references and indexes.
    ISBN 3-8055-7838-5 (hard cover : alk. paper)
    1. Eye–Diseases–Prevention.   2. Eye–Diseases–Nutritional aspects.   3.
Dietary supplements–Therapeutic use.
    [DNLM: 1. Dietary Supplements.   2. Eye Diseases–prevention & control.   3.
Drug Evaluation.   4. Nutrition. WW 140 N9756 2005]   I. Augustin, Albert J.
II. Series.
    RE48.N88 2005
    617.7′061–dc22
                                    2004021345

Bibliographic Indices. This publication is listed in bibliographic services, including Current Contents® and Index Medicus.

Drug Dosage. The authors and the publisher have exerted every effort to ensure that drug selection and dosage set forth in this text are in accord with current recommendations and practice at the time of publication. However, in view of ongoing research, changes in government regulations, and the constant flow of information relating to drug therapy and drug reactions, the reader is urged to check the package insert for each drug for any change in indications and dosage and for added warnings and precautions. This is particularly important when the recommended agent is a new and/or infrequently employed drug.

© Copyright 2005 by S. Karger AG, P.O. Box, CH–4009 Basel (Switzerland)
www.karger.com
Printed in Switzerland on acid-free paper by Reinhardt Druck, Basel
ISSN 0250–3751
ISBN 3–8055–7838–5

Contents

....

List of Contributors

E.F. Elstner, Prof. Dr.
TUM Weihenstephan, Center of Life and Food
Sciences
Lehrstuhl für Phytopathologie
Am Hochanger 2
DE-85350 Freising (Germany)

Leopold Flohé, Prof. Dr.
MOLISA GmbH
Universitätsplatz 2
DE-39106 Magdeburg (Germany)
Tel. +49 331 7480950
E-Mail l-flohe@t-online.de

K. M. Janisch, Dr.
TUM Weihenstephan, Center of Life and Food
Sciences
Lehrstuhl für Phytopathologie
Am Hochanger 2
DE-85350 Freising (Germany)

Carsten H. Meyer, Dr. med.
Department of Ophthalmology
Philipps University of Marburg
Robert-Koch-Strasse 4
DE-35037 Marburg (Germany)
Tel. +49 6421 2862616
Fax +49 6421 2865678
E-Mail meyer_eye@yahoo.com

J. Milde, Dr.
TUM Weihenstephan, Center of Life and Food
Sciences
Lehrstuhl für Phytopathologie
Am Hochanger 2
DE-85350 Freising (Germany)

H. Schempp, Dr.
TUM Weihenstephan, Center of Life and Food
Sciences
Lehrstuhl für Phytopathologie
Am Hochanger 2
DE-85350 Freising (Germany)

Ursula Schmidt-Erfurth, Prof. Dr. med.
Universitätsklinik der Augenheilkunde und
Optometrie
Währingergürtel 18-20
AT-1090 Vienna (Austria)
Tel. +43 140 4007930
Fax +43 140 4007932
E-Mail ursula.schmidt-erfurth@akhwien.at

Peter Schreier, Prof. Dr.
Lehrstuhl für Lebensmittelchemie
Universität Würzburg
Am Hubland
DE-97074 Würzburg (Germany)
E-Mail schreier@pzlc.uni-wuerzburg.de

Walter Sekundo, Priv.-Doz. Dr. med.
Department of Ophthalmology
Philipps University of Marburg
Robert-Koch-Strasse 4
DE-35037 Marburg (Germany)
Tel. +49 6421 2862643
Fax +49 6421 2865678
E-Mail sekundo@mailer.uni-marburg.de

W. Stahl, Prof. Dr.
Heinrich Heine University Düsseldorf
Institute of Biochemistry and Molecular Biology I
PO Box 101007
DE-40001 Düsseldorf (Germany)
Tel. +49 211 8112711
Fax +49 211 8113029
E-Mail wilhelm.stahl@uni-duesseldorf.de

....
Preface

In the last 10 years, there has been growing interest in antioxidants and food supplements. Many diseases of the human body, especially those of the eye, are initiated by oxidative metabolites leading to oxidative tissue damage. One of the most important agents contributing to oxidative damage of eye tissue is the physiological stimulus – light. These effects were studied extensively in the early 1980s. However, successful procedures for lens removal and replacement by intraocular lenses have reduced the enthusiasm for further research. New approaches for the treatment of age-related macular degeneration (AMD) have resulted in an overwhelming increase in the interest to continue with research in the field of AMD and related diseases. This was further enhanced by the finding that the lens can protect the macula by filtering the high-energy portion of the visible spectrum of light. In addition, we have learned that numerous diseases of the retina are mediated by oxidative tissue damage. This damage can be initiated and propagated by light- and/or inflammatory-mediated mechanisms as in AMD or by the result of oxidative metabolites of another origin. The generation of advanced glycation end products and the consecutive propagation of oxidative processes play an important role in the pathogenesis of proliferative diabetic retinopathy. Interestingly, there is an association between the generation of oxidative metabolites and inflammatory reaction and expression of growth factors such as VEGF, with cause and effect being taken into consideration.

In the aging organism, several antioxidative protective mechanisms are reduced in both their function and concentration. Consequentially, the pharmaceutical industry has put an overwhelming amount of food supplements on the market, which is confusing for physician and consumer alike. In addition, we know that the application of antioxidative agents can lead to a prooxidative or counter-reaction of those substances and disturb the balance of the antioxidative network. Fortunately, basic research findings have enhanced our knowledge of free radical mediated processes. These findings are an important contribution when making recommendations for which of antioxidative substances should be given.

This textbook is divided into two major sections: (1) basic research focusing on the major compounds of nutrition and food supplements, and (2) clinical research providing the latest information on the results of important clinical studies. The first section gives further insight into the mechanisms of action of major substances relevant to antioxidants and food supplements in relation to eye diseases. Recommendations for maximum consumption of the respective substances are given. The consequence and relevance of one of the most important trace elements – selenium – is discussed in a separate section. This is also true for vitamins E and C as well as for lutein and zeaxanthin, the physiological macular pigment. The second section focuses on both anterior and posterior segment diseases which might be influenced by food supplementation and/or antioxidants. The latest relevant studies for daily clinical work are discussed. In addition, the oxidative pathomechanisms of the most important disease processes are explained.

It is therefore hoped that this textbook, intended for clinicians and basic vision scientists, will enhance the interest in oxidative processes in eye diseases and increase research activity, especially in eye diseases leading to blindness such as diabetic retinopathy and AMD.

Albert J. Augustin
Karlsruhe

Augustin A (ed): Nutrition and the Eye.
Dev Ophthalmol. Basel, Karger, 2005, vol 38, pp 1–58

··············

Chemopreventive Compounds in the Diet

Peter Schreier

Lehrstuhl für Lebensmittelchemie, Universität Würzburg, Würzburg, Germany

Abstract

The actual knowledge about food sources, functions and nutrient-nutrient interactions, deficiencies as well as chemopreventive properties including risk/benefit considerations of vitamins C and E, selected carotenoids and flavonoids, as well as copper, zinc and selenium, and α-lipoic is reviewed.

Evidence from both epidemiological and experimental observations has shown that the high consumption of fruits and vegetables may help to prevent diseases in humans. Because of their well-documented properties, several vitamins, polyphenols and flavonoids, as well as trace elements have found particular attention as potential chemopreventive agents in our diet.

In the following, the actual knowledge about sources, functions and interactions, deficiencies as well as chemopreventive properties of vitamins C and E, carotenoids, flavonoids, copper, zinc and selenium, as well as α-lipoic acid is summarized.

Vitamin C

Vitamin C, *L*-ascorbic acid (*R*-5[(S)-1,2-dihydroxymethyl]-3,4-dihydroxy-5H-furan-2-one), is a water-soluble vitamin. Unlike most mammals, humans do not have the ability to make their own vitamin C. Therefore, we must obtain vitamin C through our diet.

Food Sources

As shown in table 1, different fruits and vegetables vary in their vitamin C content. In several foods, such as cabbage, *L*-ascorbic acid is found in form of

Table 1. Vitamin C in selected foods

Food (raw)	Vitamin C, mg/100 g
Meat products	
Beef, pork, fish	0–2
Liver, kidney	10–40
Milk	
Cow	1–2
Human	3–6
Vegetables	
Brussels sprouts	90–150
Carrot	5–10
Oat, rye, wheat	0
Kale	120–180
Potato	10–30
Rhubarb	10
Spinach	50–90
Tomato	20–35
Fruits	
Acerola	1,300
Apple	10–30
Banana	10
Citrus fruits	40–50
Guava	300
Sea buckthorn	160–800
Strawberry	40–90

L-Ascorbic acid   Dehydro-L-ascorbic acid

Ascorbigen A   Ascorbigen B

ascorbigen A (B) that is split into *L*-ascorbic acid in the course of cooking. However, thermal treatment usually destroys a considerable part of vitamin C.

Function

There are many functions related to vitamin C. It is required for the synthesis of collagen and also plays an important role in the syntheses of norepinephrine and carnitine [1]. Recent research also suggests that vitamin C is involved in the metabolism of cholesterol to bile acids [2].

Vitamin C is a highly effective antioxidant and can protect proteins, lipids, carbohydrates, and nucleic acids from damage by free radicals and reactive oxygen species that can be generated during normal metabolism as well as through exposure to toxins and pollutants. Vitamin C may also be able to regenerate other antioxidants, such as vitamin E [1].

However, vitamin C can interact in vitro with some free metal ions to produce potentially damaging free radicals. Although free metal ions are not generally found under physiological conditions, the idea that high doses of vitamin C might be able to promote oxidative damage in vivo has received a great deal of attention. A recent review found no sufficient scientific evidence that vitamin C promotes oxidative damage under physiological conditions or in humans [3].

Deficiency

Severe vitamin C deficiency has been known for many centuries as the disease scurvy. By the late 1700s the British navy was aware that scurvy could be cured by eating oranges or lemons, even though vitamin C would not be isolated until the early 1930s. At present, scurvy is rare in developed countries because it can be prevented by as little as 10 mg of vitamin C daily.

In the USA, the recommended dietary allowance (RDA) for vitamin C was revised upward from 60 mg daily for men and women (table 2). The recommended intake for smokers is 35 mg/day higher than for non-smokers, because smokers are under increased oxidative stress and generally have lower blood levels of vitamin C [4].

Disease Prevention

Much of the information regarding vitamin C and the prevention of chronic disease is based on prospective studies in which vitamin C intake is assessed in large numbers of people who are followed over time to determine whether they develop specific chronic diseases.

Cardiovascular Diseases

Until recently, the results of most studies indicated that low or deficient intakes of vitamin C were associated with an increased risk of cardiovascular

Table 2. RDA for vitamin C in the USA

Life stage	Age	Males, mg/day	Females, mg/day
Infants	0–6 months	40	40
Infants	7–12 months	50	50
Children	1–3 years	15	15
Children	4–8 years	25	25
Children	9–13 years	45	45
Adolescents	14–18 years	75	65
Adults	19 years and older	90	75
Smokers	19 years and older	125	110
Pregnancy	18 years and younger	–	80
Pregnancy	19 years and older	–	85
Breast-feeding	18 years and younger	–	115
Breast-feeding	19 years and older	–	120

diseases and that modest dietary intake of about 100 mg/day was sufficient to reduce the risk among non-smoking men and women [1]. Several studies failed to find significant reductions in the risk of coronary heart disease (CHD) among vitamin C supplement users [5, 6]. The First National Health and Nutrition Examination Study (NHANES I) [7] found that the risk of death from cardiovascular diseases was 42% lower in men and 25% lower in women who consumed >50 mg/day of dietary vitamin C [8]. Recent results from the Nurses' Health Study based on the follow-up of more than 85,000 women over 16 years also suggest that higher vitamin C intakes may be cardio-protective [9]. In this study, vitamin C intakes of >300 mg/day from diet plus supplements or supplements were associated with a 27–28% reduction in CHD risk. However, in those women who did not take vitamin C supplements, dietary vitamin C intake was not significantly associated with CHD risk. This finding is inconsistent with data from numerous other prospective cohort studies that found inverse associations between dietary vitamin C intake of vitamin C plasma levels and CHD risk [1, 10]. Data from the National Institutes of Health (NIH) indicated that plasma and circulating cells in healthy, young subjects became fully saturated with vitamin C at a dose of 400 mg/day [11]. The results of the NHANES I study and the Nurses' Health Study suggest that maximum reduction of cardiovascular disease risk may require vitamin C intakes high enough to saturate plasma and circulating cells, and thus the vitamin C body pool [12].

Cancer

From a number of studies it has been concluded that increased consumption of fresh fruits and vegetables is associated with a reduced risk of several types of cancer [13]. Such studies are the basis for dietary guidelines which recommend at least 5 servings of fruits and vegetables per day.

In several case-control studies the role of vitamin C in cancer prevention has been investigated. Most have shown that higher intakes of vitamin C are associated with a decreased incidence of several cancers. In general, prospective studies in which the lowest intake group consumed >86 mg of vitamin C daily have not found differences in cancer risk, while studies finding significant cancer risk reductions found them in people consuming at least 80–110 mg of vitamin C daily [1].

Although most large prospective studies found no association between breast cancer and vitamin C intake, two recent studies described dietary vitamin C intake to be inversely associated with breast cancer risk in certain subgroups [14]. In the Swedish Mammography Cohort, women who consumed an average of 110 mg/day of vitamin C had a 39% lower risk of breast cancer compared to women who consumed an average of 31 mg/day [15]. A number of observational studies have found increased dietary vitamin C intake to be associated with decreased risk of stomach cancer, and laboratory experiments indicate that vitamin C inhibits the formation of carcinogenic compounds in stomach. Infection with *Heliobacter pylori* is known to increase the risk of stomach cancer and also appears to lower the vitamin C content of stomach secretions. Although two intervention studies did not find a decrease in the occurrence of stomach cancer with vitamin C supplementation [4], more recent research suggests that vitamin C supplementation may be a useful addition to standard *H. pylori* eradication therapy in reducing the risk of gastric cancer [16].

Cataracts

Decreased vitamin C levels in the lens of the eye have been associated with increased severity of cataracts in humans. Some, but not all studies have observed increased dietary vitamin C intake [17] and increased blood levels of vitamin C [18] to be associated with decreased risk of cataracts. Those studies that have found a relationship suggest that vitamin C intake may have to be >300 mg/day for a number of years before a protective effect can be detected [1]. Recently, a 7-year controlled intervention trial of a daily antioxidant supplement containing 500 mg of vitamin C, 400 IU of vitamin E, and 15 mg of β-carotene in 4,629 men and women found no difference between the antioxidant combination and a placebo on the development and progression of age-related cataracts [19]. Therefore, the relationship between vitamin C intake and the development of cataracts requires further clarification.

Table 3. Tocopherols and tocotrienols

	R¹	R²	R³	Configuration	Optical activity
α-T.	CH_3	CH_3	CH_3	$2R,4'R,8'R$	$[\alpha]^{25}D$
β-T.	CH_3	H	CH_3	$2R,4'R,8'R$	$[\alpha]^{20}D$ + 6.37°
γ-T.	H	CH_3	CH_3	$2R,4'R,8'R$ ($\pm$)-Form	$[\alpha]^{25}456 - 2.4°$ (C_2H_5OH)
δ-T	H	H	CH_3	$2R,4'R,8'R$	
ζ-T.	CH_3	CH_3	H	$2R,4'R,8'R$	
η-T.	H	CH_3	H	$2R,4'R,8'R$	
α	CH_3	CH_3	CH_3	(E,E) R-(E,E)	$[\alpha]^{25}D$ − 5.7° ($CHCl_3$)
β⁻(ε-T.)	CH_3	H	CH_3	R-(E,E)	
γ	H	CH_3	CH_3	R-(E,E)	
α	CH_3	CH_3	CH_3	$3'R,7'R,11'R$	
α	CH_3	CH_3	CH_3	$3'R,7'R,11'R$	

Safety

A number of possible problems with very large doses of vitamin C have been suggested, mainly based on in vitro experiments or isolated case reports, including: genetic mutations, birth defects, cancer, atherosclerosis, kidney stones, 'rebound' scurvy, increased oxidative stress, excess iron absorption, and vitamin B_{12} deficiency. However, none of these adverse health effects have been confirmed, and there is no reliable scientific evidence that large amounts of vitamin C are toxic or detrimental to health. With the latest RDA published in 2000, a tolerable upper intake level (UL) of 2 g daily was recommended [4].

Vitamin E

The term vitamin E comprises a family of several antioxidants, i.e. tocopherols and tocotrienols (cf. formula and table 3). α-Tocopherol is the only form of vitamin E found in the largest quantities in the blood and tissue [20]. As α-tocopherol is the form of vitamin E that appears to have the greatest nutritional significance, it will be the primary topic of the following discussion.

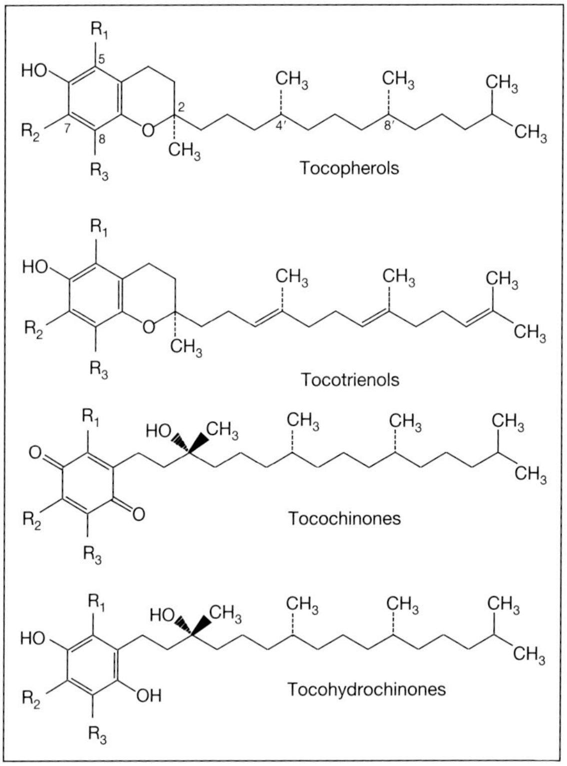

HO

R₁

R₂

R₃

CH₃

CH₃   CH₃   CH₃

CH₃

Tocopherols

HO

R₁

R₂

R₃

CH₃

CH₃   CH₃   CH₃

CH₃

Tocotrienols

O

R₁

R₂

R₃

O

HO CH₃

CH₃   CH₃   CH₃

CH₃

Tocochinones

HO

R₁

R₂

R₃

OH

HO CH₃

CH₃   CH₃   CH₃

CH₃

Tocohydrochinones

Food Sources

Major sources of α-tocopherol in the diet include vegetable oils (olive, sunflower, safflower oils) nuts, whole grains and green leafy vegetables. All forms of vitamin E occur naturally in foods, but in varying amounts. Selected examples are given in table 4.

Function

α-Tocopherol

The main function of α-tocopherol in humans appears to be that of an antioxidant. The fat-soluble vitamin is suited to intercepting free radicals and preventing a chain reaction of lipid destruction. Aside from maintaining the integrity of cell membranes throughout the body, α-tocopherol also protects the lipids in low density lipoproteins (LDL) from oxidation.

Several other functions of α-tocopherol have been described. It is known to inhibit the activity of protein kinase C, an important cell signaling molecule, as well as to affect the expression and activity of immune and inflammatory

Table 4. Amounts of vitamin E in selected foods (mg/100 g)

	α-Tocopherol	β-Tocopherol	γ-Tocopherol	Others
Milk, cow pasteurized	0.08	–	–	–
Milk, human	0.5	–	–	–
Egg	0.7	–	0.4	–
Butter	2.2	–	–	–
Margarine	14.0	–	–	–
Olive oil	11.9	0.1	0.6	–
Maize oil	25.0	0.65	55.8	2.5
Wheat germ oil	192.0	50.8	30.4	6.8
Spinach	1.6	–	0.1	0.8
Tomato	0.8	–	0.1	–
Wheat, whole grain	1.0	0.4	–	2.9

cells. In addition, α-tocopherol has been shown to inhibit platelet aggregation and to enhance vasodilation [21, 22].

γ-Tocopherol
The function of γ-tocopherol in humans is still unclear. Its blood levels are generally 10 times lower than those of α-tocopherol. Limited in vitro and animal tests indicate that γ-tocopherol or its metabolites may play a role in the protection of the body from damage of free radicals [23, 24], however these effects have not been demonstrated convincingly in humans.

In one recent prospective study increased plasma γ-tocopherol levels were associated with a significantly reduced risk of developing prostate cancer, while significant protective associations for increased levels of plasma α-tocopherol were found only when γ-tocopherol levels were also high [25]. These limited findings, in addition to the fact that taking α-tocopherol supplements may lower γ-tocopherol levels in blood, have activated the interest for additional research on the effects of dietary and supplemental γ-tocopherol on health [26].

Deficiency
Vitamin E deficiency has been observed in individuals with malnutrition, genetic defects affecting the α-tocopherol transfer protein and fat malabsorption syndromes. Severe vitamin E deficiency results mainly in neurological symptoms (ataxia and peripheral neuropathy), myopathy, and pigmented retinopathy. For this reason, people who develop peripheral neuropathy, ataxia or retinitis pigmentosa should be screened for vitamin E deficiency [27].

Table 5. RDA for (*RRR*)-α-tocopherol in the USA

Life stage	Age	Males mg/day	Females mg/day
Infants	0–6 months	4	4
Infants	7–12 months	5	5
Children	1–3 years	6	6
Children	4–8 years	7	7
Children	9–13 years	11	11
Adolescents	14–18 years	15	15
Adults	19 years and older	15	15
Pregnancy	All ages	–	15
Breast-feeding	All ages	–	19

Although true vitamin E deficiency is rare, suboptimal intake of vitamin E is quite common. The National Health and Nutrition Examination Survey III (NHANES III) investigated the dietary intake and blood levels of α-tocopherol in 16,295 multi-ethnic adults. 27% of white participants, 41% of African-Americans, 28% of Mexican-Americans and 32% of the other participants were found to have blood levels of α-tocopherol $<20\,\mu$mol/l, a value chosen because the literature suggests an increased risk for cardiovascular disease below this level [28].

The RDA for vitamin E was previously 8 mg/day for women and 10 mg/day for men, but it was revised in 2000 [21] (table 5).

Disease Prevention

Cardiovascular Diseases

The results of large observational studies suggest that increased vitamin E consumption is associated with decreased risk of myocardial infarction or death from heart disease in both men and women. Each study was a prospective study which measured vitamin E consumption in presumably healthy people and followed them for number of years to determine how many of them were diagnosed with, or died as a result of heart disease. In two of the studies, those individuals who consumed >7 mg of α-tocopherol in food were only approximately 35% as likely to die from heart disease as those who consumed <3–5 mg of α-tocopherol [29, 30]. Two other large studies found a significant reduction in the risk of heart disease only in those women and men who consumed α-tocopherol supplements of at least 67 mg of (*RRR*)-α-tocopherol daily [31, 32]. Recently, several studies have observed plasma or red blood cell levels of α-tocopherol to be inversely associated with the presence or severity supplements in patients with heart disease have not shown vitamin E to be effective in preventing heart attacks or death [33, 34].

Cancer

Several large prospective studies have failed to find significant associations between α-tocopherol intake and the incidence of lung cancer or breast cancer [21]. A placebo-controlled intervention study designed to look at the effect of α-tocopherol supplementation on lung cancer in smokers found a 34% reduction in the incidence of prostate cancer in smokers given supplements of 50 mg of synthetic α-tocopherol (equivalent to 25 mg of (RRR)-α-tocopherol) daily [35].

Cataracts

To date, ten observational studies have examined the association between vitamin E consumption and the incidence and severity of cataracts. Of these studies, five found increased vitamin E intake to be associated with protection from cataracts, while five reported no association [36, 37]. A recent intervention trial of a daily antioxidant supplement containing 500 mg of vitamin C, 400 IU of vitamin E, and 15 mg of β-carotene in 4,629 men and women found that the antioxidant supplement was not different than a placebo in its effects on the development and progression of age-related cataracts over a 7-year period [38]. Another invention trial found that a daily supplement of 50 mg of synthetic α-tocopherol daily (equivalent to 25 mg (RRR)-α-tocopherol) did not alter the incidence of cataract surgery in male smokers [39]. Thus, the relationship between vitamin E intake and the development of cataracts requires further clarification.

Immune Function

α-Tocopherol has been shown to enhance specific aspects of the immune response. For instance, 200 mg of synthetic α-tocopherol (equivalent to 100 mg (RRR)-α-tocopherol) daily for several months increased the formation of antibodies in response to hepatitis B vaccine and tetanus vaccine in elderly adults [40]. Whether α-tocopherol-associated enhancements in the immune response actually translate to increased resistance to infections in older adults remains to be determined [41].

Safety

Few side effects have been noted in adults taking supplements of <2,000 mg of α-tocopherol daily ([RRR]- or racemic α-tocopherol). However, results of long-term α-tocopherol supplementation have not been studied. In addition

Table 6. Tolerable UL for α-tocopherol in the USA

Age group	UL, mg/day (IU/day) *d*-α-tocopherol
Infants 0–12 months	Not possible to establish
Children 1–3 years	200 (300)
Children 4–8 years	300 (450)
Children 9–13 years	600 (900)
Adolescents 14–18 years	800 (1,200)
Adults 19 years and older	1,000 (1,500)

to setting the new RDA for α-tocopherol in 2000, an upper limit (UL) for α-tocopherol was given (table 6). An UL of 1,000 mg daily of α-tocopherol of any form would be the highest dose unlikely to result in potential hemorrhage in adults [21].

Copper

Copper (Cu) is an essential trace element for animals and humans. In the body, copper shifts between the cuprous (Cu^{1+}) and the cupric (Cu^{2+}) forms, though the majority of the body's copper is in the Cu^{2+} form. The ability of copper to easily accept and donate electrons explains its important role in oxidoreductions and scavenging of free radicals [42].

Food Sources
Copper is found in a wide variety of foods and is most plentiful in organ meats, shellfish, nuts, and seeds. Wheat bran cereals and whole grain products are also good sources of copper. According to national surveys, the average dietary intake of copper in the USA is approximately 1.0–1.1 mg/day for adult women and 1.2–1.6 mg/day for adult men.

Function
Copper is a critical functional component of a number of essential enzymes. Some of the physiological functions known to be copper-dependent are discussed below.

Cytochrome *c* oxidase plays an essential role in cellular energy production. By catalyzing the reduction of molecular oxygen to water, an electrical gradient is generated used by the mitochondria to create ATP [43].

Another cuproenzyme, lysyl oxidase, is required for the cross-linking of collagen and elastin. The action of lysyl oxidase helps to maintain the integrity of connective tissue in the heart and blood vessels and plays a role in bone formation [44].

Ceruloplasmin (ferroxidase I) and ferroxidase II have the capacity to oxidize ferrous iron (Fe^{2+}) to ferric iron (Fe^{3+}), the form of iron that can be loaded onto the protein transferring for transport to the site of red blood cell formation. Although the ferroxidase activity of these two cuproenzymes has not yet been proven to be physiologically significant, the fact that iron mobilization from storage sites is impaired in copper deficiency supports their role in iron metabolism [45].

A number of reactions essential to normal function of the brain and nervous system are catalyzed by cuproenzymes. Monoamine oxidase (MAO) plays a role in the metabolism of norepinephrine, epinephrine, and dopamine. MAO also functions in the degradation of serotonin which is the basis for the use of MAO inhibitors as antidepressants [46]. The myelin sheath is made of phospholipids whose synthesis depends on cytochrome c oxidase activity [44].

In addition, tyrosinase is required for the formation of the pigment melanin. Melanin is formed in the melanocytes and plays a role in the pigmentation of the hair, skin and eyes [44].

Furthermore, copper has several antioxidant functions. Superoxide dismutase (SOD) is operative as an antioxidant by catalyzing the conversion of superoxide radicals to hydrogen peroxide which can subsequently be reduced to water enzymatically [47]. Two forms of SOD contain copper, i.e. (i) copper/zinc SOD is found within most cells of the body, including red blood cells, and (ii) extracellular SOD is a copper-containing enzyme found in high levels in the lung and low levels in blood plasma [44].

Finally, copper-dependent transcription factors regulate transcription of specific genes. Thus, cellular copper levels may affect the synthesis of proteins by enhancing or inhibiting the transcription of specific genes. They include genes for Cu/Zn SOD, catalase, and proteins related to the cellular storage of copper [43].

Nutrient-Nutrient Interactions

Iron. An adequate copper nutritional status appears to be necessary for normal iron metabolism and red blood cell formation. Iron has been found to accumulate in the livers of copper-deficient animals, indicating that copper (probably in the form of ceruloplasmin) is required for iron transport to the bone marrow for red blood cell formation [44]. Infants fed a high iron formula absorbed less copper than infants fed a low iron formula, suggesting that high iron intakes may interfere with copper absorption in infants [46].

Zinc. High supplemental zinc intakes of ≥50 mg/day for extended periods of time may result in copper deficiency. High dietary zinc increases the synthesis of metallothionein, which binds certain metals and prevents their absorption by trapping them in intestinal cells. Metallothionein has a stronger affinity for copper than zinc, so, high levels of metallothionein induced by excess zinc cause a decrease in intestinal copper absorption. High copper intakes have not been found to affect zinc nutritional status [42, 47].

Vitamin C. Although vitamin C supplements have produced copper deficiency in laboratory animals, the effect of vitamin C supplements on copper nutritional status in humans is less clear. In one human study, vitamin C supplementation of 1,500 mg/day for 2 months resulted in a significant decline in ceruloplasmin oxidase activity [48]. In another one, supplements of 605 mg of vitamin C/day for 3 weeks also resulted in decreased ceruloplasmin oxidase activity, although copper absorption did not decline. Neither of these studies found vitamin C supplementation to adversely affect copper nutritional status.

Deficiency

Clinically evident copper deficiency is uncommon. Serum copper and ceruloplasmin levels may fall to 30% of normal in cases of severe copper deficiency. One of the most common clinical signs of copper deficiency is an anemia that is unresponsive to iron therapy but corrected by copper supplementation. The anemia is thought to result from defective iron mobilization due to decreased ceruloplasmin activity. Copper deficiency may also result in neutropenia, a condition that may be accompanied by increased susceptibility to infection. Abnormalities of bone development related to copper deficiency are most common in copper-deficient low-birth-weight infants and young children. Less common features of copper deficiency may include loss of pigmentation, neurological symptoms, and impaired growth [43, 44].

Individuals at Risk of Deficiency

High-risk individuals include premature infants, especially those with low birth weight, infants with prolonged diarrhea, infants and children recovering from malnutrition, individuals with malabsorption syndromes, including celiac disease, sprue and short bowel syndrome due to surgical removal of a large portion of the intestine. Individuals receiving intravenous total parental nutrition or other restricted diets may also require supplementation with copper and other trace elements [43, 44]. Recent research indicates that cystic fibrosis patients may also be at increased risk of copper insufficiency [49].

Table 7. RDA for copper in the USA

Life stage	Age	Males µg/day	Females µg/day
Infants	0–6 months	200	200
Infants	7–12 months	220	220
Children	1–3 years	340	340
Children	4–8 years	440	440
Children	9–13 years	700	700
Adolescents	14–18 years	890	890
Adults	19 years and older	900	900
Pregnancy	All ages	–	1,000
Breast-feeding	All ages	–	1,300

A variety of indicators were used to establish to RDA for copper, including plasma copper concentration, serum ceruloplasmin activity, SOD activity in red blood cells, and platelet copper concentration [46]. The RDA for copper reflects the results of depletion-repletion studies and is based on the prevention of deficiency (table 7).

Disease Prevention
Cardiovascular Diseases
While severe copper deficiency results in cardiomyopathy in some animal species, the pathology differs from atherosclerotic cardiovascular diseases prevalent in humans. Studies in humans have produced inconsistent results and their interpretation is hindered by the lack of a reliable marker of copper nutritional status. Outside the body, free copper is known to be a pro-oxidant and is frequently used to produce LDL oxidation. Recently, ceruloplasmin has been found to stimulate LDL oxidation in vitro [50], however, there is little evidence that copper or ceruloplasmin promotes LDL oxidation in vivo. In addition, SOD and ceruloplasmin are known to have antioxidant properties leading to assume that copper deficiency rather than excess copper increases the risk of cardiovascular diseases [51].

Several epidemiologic studies have found increased serum copper levels to be associated with increased risk of cardiovascular disease. A recent prospective study in the USA examined serum copper levels in more than 4,400 men and women at the age of 30 years and older [52]. During the following 16 years, 151 participants died from CHD. After adjusting for other risk factors of heart disease, those with serum copper levels in the two highest quartiles had a significantly greater risk of dying from CHD. Three other case-control studies conducted in Europe had similar findings. Serum copper largely reflects serum ceruloplasmin and is not a sensitive indicator of copper nutritional status. Serum ceruloplasmin levels are known to increase by 50% or more under certain conditions of physical stress, such as trauma, inflammation, or disease. Because over 90% of serum copper is carried in ceruloplasmin, elevated serum copper may simply be a marker of the inflammation that accompanies atherosclerosis.

In contrast to the serum copper findings, two autopsy studies found copper levels in heart muscle to be lower in patients who died of CHD than those who died of other causes [53]. Additionally, the copper content of white blood cells has been positively correlated with the degree of patency of coronary arteries in CHD patients [54, 55] and patients with a history of myocardial infarction had lower concentrations of extracellular SOD than those without such a history [56].

While in small human studies adverse changes in blood cholesterol levels, including increased total and LDL-cholesterol levels and decreased HDL-cholesterol levels [57] were found, other studies have not confirmed those results [58]. Copper supplementation of 2–3 mg/day for 4–6 weeks did not result in clinically significant changes in cholesterol levels [51, 59]. Recent research has also failed to find evidence that increased copper intake increases oxidative stress. In a multicenter placebo-controlled study, copper supplementation of 3 and 6 mg/day for 6 weeks did not result in increased susceptibility of LDL ex vivo oxidation [60]. Moreover, in vitro the oxidizability of red blood cells decreased [61], indicating that relatively high intakes of copper do not increase the susceptibility of LDL or red blood cells to oxidation.

Osteoporosis

Osteoporosis has been observed in infants and adults with severe copper deficiency, but it is not clear whether marginal copper deficiency contributes to osteoporosis. Research regarding the role of copper nutritional status in age-related osteoporosis is limited. Serum copper levels of 46 elderly patients with hip fractures were reported to be significantly lower than matched controls [62]. A small study in premenopausal women who consumed an average of 1 mg of dietary copper daily, reported a decreased loss of bone mineral density from the lumbar spine after copper supplementation of 3 mg/day for 2 years [63]. Marginal copper intake of 0. 7 mg/day for 6 weeks significantly increased a measurement of bone resorption (breakdown) in healthy adult males [64]. However, copper supplementation of 3–6 mg/day for 6 weeks had no effect on biochemical markers of bone resorption or bone formation in a study of healthy adult men and women [65].

Immune Function

The exact mechanism of the action of copper on the immune system function is not yet known. Adverse effects in insufficient copper appear most

Table 8. Tolerable UL for copper in the USA

Age group	UL, mg/day
Infants 0–12 months	Not possible to establish
Children 1–3 years	1
Children 4–8 years	3
Children 9–13 years	5
Adolescents 14–18 years	8
Adults 19 years and older	10

pronounced in infants. Infants with Menkes disease, a genetic disorder that results in severe copper deficiency, suffer from frequent and severe infections [66, 67]. In a study of 11 malnourished infants with evidence of copper deficiency, the ability of certain white blood cells to engulf pathogens increased significantly after 1 month of copper supplementation [68]. More recently, 11 men on a low-copper diet (0.66 mg copper/day for 24 days and 0.38 mg/day for another 40 days) showed a decreased proliferation response when mononuclear cells were presented with an immune challenge in cell culture [69]. While severe copper deficiency has adverse effects on immune function, the effects of marginal copper insufficiency in humans are not clear.

Safety

Copper toxicity is rare in the general population. Acute copper poisoning has occurred through the contamination of beverages by storage in copper-containing containers as well as from contaminated water supplies [70]. In the USA, the health-based guideline for a maximum water copper concentration of 1.3 mg/l is enforced by the Environmental Protection Agency (EPA) [71]. Symptoms of acute copper toxicity include abdominal pain, nausea, vomiting, and diarrhea which help to prevent additional ingestion and absorption of copper. More serious signs of acute copper toxicity include severe liver damage, kidney failure, coma, and death.

As to healthy individuals, doses of up to 10 mg daily have not resulted in liver damage. For this reason, the UL for copper was set at 10 mg/day from food and supplements [46] (table 8).

Selenium

Selenium is a trace element that is essential in small amounts but can be toxic in larger quantities. Humans and animals require selenium for the function of a number of selenoproteins. During their synthesis, selenocysteine is incorporated into a very specific location in the amino acid sequence in order to form a functional protein. Unlike animals, plants do not appear to require selenium. However, when selenium is present in the soil, plants incorporate it non-specifically into compounds that usually contain sulfur [72, 73].

Food Sources

The richest food sources of selenium are organ meats and seafood, followed by muscle meats. In general, there is a wide variation in the selenium content of plants and grains, as the incorporation of selenium into plant proteins is dependent only on soil selenium content. Brazil nuts grown in areas of Brazil with selenium-rich soil may provide >100 µg of selenium in one nut, while those grown in selenium-poor soil may contain 10 times less [74]. In the USA, grains are a good source of selenium, but fruits and vegetables tend to be relatively poor sources. In general, drinking-water is not a significant source of selenium. The average dietary intake of adults in the USA has been found to range from about 80 to 110 µg/day.

Function

At least 11 selenoproteins have been characterized and there is evidence that additional selenoproteins exist.

Glutathione Peroxidases. Four selenium-containing glutathione peroxidases (GPx) have been identified, i.e. (i) cellular or classical GPx, (ii) plasma or extracellular GPx, (iii) phospholipid hydroperoxide GPx, and (iv) gastrointestinal GPx [75]. Although each GPx is a distinct selenoprotein, they are all antioxidant enzymes that reduce potentially damaging reactive oxygen species, such as hydrogen peroxide and lipid hydroperoxides to harmless products like water and alcohols by coupling their reduction with the oxidation of glutathione.

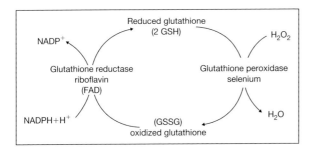

Sperm mitochondrial capsule selenoprotein was once thought to be a distinct selenoprotein, but now appears to be phospholipid hydroperoxide GPx [76].

Thioredoxin Reductase. Thioredoxin reductase participates in the regeneration of several antioxidant systems, possibly including vitamin C. Maintenance of thioredoxin in a reduced form by thioredoxin reductase is important for regulating cell growth and viability [75, 77].

Iodothyronine Deiodinases. The thyroid gland releases very small amounts of biologically active thyroid hormone (triiodothyronine, T_3) and larger amounts of an inactive form of thyroid hormone (thyroxine, T_4) into the circulation. Most of the biologically active T_3 in the circulation and inside cells is created by the removal of one iodine atom from T_4 in a reaction catalyzed by selenium-dependent iodothyronine deiodinase enzymes. Through their actions on T_3, T_4 and other thyroid hormone metabolites, three different selenium-dependent iodothyronine deiodinases (types I, II, and III) can both activate and inactivate thyroid hormone, making selenium an essential element for normal development, growth, and metabolism through the regulation of thyroid hormones [75, 78].

Selenoprotein P and W. Selenoprotein P is found in plasma and is also associated with vascular endothelial cells. Although the function of selenoprotein P has not been clarified to date, it has been suggested to function as a transport protein as well as an antioxidant capable of protecting endothelial cells from peroxynitrite damage [75, 79]. Selenoprotein W is found in the muscle. Although its function is presently unknown, it is thought to play a role in muscle metabolism [75].

Selenophosphate Synthetase. Incorporation of selenocysteine into selenoproteins is directed by the genetic code and requires the enzyme selenophosphate synthetase. A selenoprotein itself, selenophosphate synthetase catalyzes the synthesis of monoselenium phosphate, a precursor of selenocysteine which is required for the synthesis of selenoproteins [73, 75].

Nutrient–Nutrient Interactions

Antioxidant Nutrients. As an integral part of the GPx and thioredoxin reductase, selenium probably interacts with every nutrient that affects the pro-oxidant/antioxidant balance of the cell. Other minerals that are critical components of antioxidant enzymes include copper, zinc, and iron. Selenium as GPx also appears to support the activity of α-tocopherol in limiting the oxidation of lipids. Animal studies indicate that selenium and α-tocopherol tend to spare one another and that selenium can prevent some of the damage resulting from α-tocopherol deficiency in models of oxidative stress. Thioredoxin reductase also maintains the antioxidant function of vitamin C by catalyzing its regeneration [77].

Iodine. Selenium deficiency may impair the effects of iodine deficiency. Iodine is essential for the synthesis of thyroid hormone, but the iodothyronine deiodonases are also required for the conversion of T_4 to the biologically active thyroid hormone T_3. Selenium supplementation in a small group of elderly individuals decreased plasma T_4, indicating increased deiodinase activity with increased conversion to T_3 [73].

Deficiency
Insufficient selenium intake results in decreased activity of the GPx. Even when severe, isolated selenium deficiency does not usually result in obvious clinical illness.

Individuals at Risk of Deficiency
Clinical selenium deficiency has been observed in chronically ill patients who were receiving total parenteral nutrition (TPN) without added selenium for prolonged periods of time. Muscular weakness, muscle wasting, and cardiomyopathy have been observed. TPN solutions are now supplemented with selenium to prevent such problems. People who have had a large portion of the small intestine surgically removed or those with severe gastrointestinal problems, such as Crohn's disease, are also at risk for selenium deficiency due to impaired absorption. Specialized medical diets used to treat metabolic disorders, such as phenylketonuria, are often low in selenium. Specialized diets that will be used exclusively over long periods of time should have their selenium content assessed to determine the need for selenium supplementation [72].

Keshan disease is a cardiomyopathy that affects young women and children in a selenium-deficient region of China. The incidence of Keshan disease is closely associated with very low dietary intakes of selenium and poor selenium nutritional status. Despite the strong evidence that selenium deficiency is a fundamental factor in the etiology of Keshan's disease, the seasonal and annual variation in its occurrence suggests that an infectious agent is also involved. Coxsackie virus is one of the viruses that has been isolated from Keshan patients and this virus has been found to be capable of causing an inflammation of the heart called myocarditis in selenium-deficient mice. Studies in mice indicate that oxidative stress induced by selenium deficiency results in changes in the viral genome capable of converting a relatively harmless viral strain to a myocarditis-causing strain [80, 81]. Though not proven in Keshan disease, selenium deficiency may result in a more virulent strain of virus with the potential to invade and damage the heart muscle.

Kashin-Beck disease is characterized by osteoarthritis and is associated with poor selenium status in areas of northern China, North Korea, and eastern Siberia. The disease affects children between the ages of 5 and 13 years. Unlike Keshan disease, there is little evidence that improving selenium nutritional status prevents Kashin-Beck disease. Thus, the role of selenium deficiency in the

Table 9. RDA for selenium in the USA

Life stage	Age	Males μg/day	Females μg/day
Infants	0–6 months	15	15
Infants	7–12 months	20	20
Children	1–3 years	20	20
Children	4–8 years	30	30
Children	9–13 years	40	40
Adolescents	14–18 years	55	55
Adults	19 years and older	55	55
Pregnancy	All ages	–	60
Breast-feeding	All ages	–	70

etiology of Kashin-Beck disease is less certain. A number of other causative factors have been suggested for Kashin-Beck disease, including fungal toxins in grain, iodine deficiency, and contaminated drinking-water [82].

The most recent RDA is based on the amount of dietary selenium required to maximize the activity of the antioxidant enzyme glutathione peroxidase in blood plasma [81] (table 9).

Disease Prevention
Cardiovascular Diseases

From the theory, optimizing selenoenzyme activity could decrease the risk of cardiovascular diseases by reducing lipid peroxidation and influencing the metabolism of cell signaling molecules known as prostaglandins. However, prospective studies in humans have not demonstrated strong support for the cardio-protective effects of selenium. While one study found a significant increase in illness and death from cardiovascular disease in individuals with serum selenium levels <45 μg/l compared to matched pairs >45 μg/l [83], another study using the same cut-off points for serum selenium found a significant difference only in deaths from stroke [84]. A study of middle-aged and elderly Danish men found an increased risk of cardiovascular disease in men with serum selenium levels <79 μg/l [85], but several other studies found no clear inverse association between selenium nutritional status and cardiovascular disease risk [86]. In a multicenter study in Europe, toenail selenium levels and risk of myocardial infarction (hear attack) were only associated in the center where selenium levels were the lowest [87]. While some epidemiologic evidence suggests that low levels of selenium may increase the risk of cardiovascular diseases, definitive evidence regarding the role of selenium in preventing cardiovascular diseases will require defined clinical trials.

Cancer

There is good evidence that selenium supplementation at high levels reduced the incidence of cancer in animals. More than two-thirds of over 100 published studies in 20 different animal models of spontaneous, viral, and chemically induced cancers found that selenium supplementation significantly reduced tumor incidence [88]. The evidence indicates that the methylated forms of selenium are the active species against tumors, and these methylated selenium compounds are produced at the greatest amounts with excess selenium intakes. Selenium efficiency does not appear to make animals more susceptible to develop cancerous tumors [89].

Epidemiologic Studies. A prospective study of more than 60,000 female nurses in the USA found no association between toenail selenium levels and total cancer risk [90]. In a study of Taiwanese men with chronic viral hepatitis B or C infection decreased plasma selenium concentrations were associated with an even greater risk of liver cancer [91]. A case-control study within a prospective study of over 9,000 Finnish men and women examined serum selenium levels in 95 individuals subsequently diagnosed with lung cancer and 190 matched controls [92]. Lower serum selenium levels were associated with an increased risk of lung cancer and the association was more pronounced in smokers. In this Finnish population, selenium levels were only about 60% of that common in generally observed other Western countries.

Another case-control study within a prospective study of over 5,000 male health professionals in the USA found a significant inverse relationship between toenail selenium content and the risk of prostate cancer in 181 men diagnosed with advance prostate cancer and 181 matched controls [93]. In individuals whose toenail selenium content was consistent with an average intake of 159 µg/day, the risk of advance prostate cancer was only 35% of those individuals with toenail selenium content consistent with an intake of 86 µg/day. Within a prospective study of more than 9,000 Japanese-American men, a case-control study that examined 249 confirmed cases of prostate cancer and 249 matched controls found the risk of developing prostate cancer to be 50% less in men with serum selenium levels in the highest quartile compared to those in the lowest quartile [94], while another case-control study found that men with prediagnostic plasma selenium levels in the lowest quartile were 4–5 times more likely to develop prostate cancer than those in the highest quartile [95]. In contrast, one of the largest case-control studies to date found a significant inverse association between toenail selenium and the risk of colon cancer, but no associations between toenail selenium and the risk of breast cancer or prostate cancer [96].

Human Intervention Trials. (1) Undernourished populations: An intervention trial undertaken among a general population of 130,471 individuals in five townships of Qidong, China, a high-risk area for viral hepatitis B infection and liver cancer, provided table salt enriched with sodium selenite to the population of one township (20,847 people) using the other four townships as controls. During an 8-year follow-up period, the average incidence of liver cancer was reduced by 35% in the selenium-enriched population while no reduction was found in the control populations. In a clinical trial in the same region, 226 individuals with evidence of chronic hepatitis B infection were supplemented with either 200 μg of selenium in the form of a selenium-enriched yeast tablet or a placebo yeast tablet daily. During the 4-year follow-up period, 7 out of 113 individuals supplemented with the placebo developed primary liver cancer while none of the 113 subjects supplemented with selenium developed liver cancer [97]. *(2) Well-nourished populations:* In the USA, a double-blind, placebo-controlled study of more than 13,000 older adults with a history of non-melanoma skin cancer found that supplementation with 200 μg/day of selenium-enriched yeast for an average of 7.4 years was associated with a 51% decrease in prostate cancer incidence in men [98]. The protective effect of selenium supplementation was the greatest in those men with lower baseline plasma selenium and prostate-specific antigen (PSA) levels. Surprisingly, recent results from this study indicate that selenium supplementation increased the risk of one type of skin cancer (squamous cell carcinoma) by 25% [99]. Although selenium supplementation shows promise for the prevention of prostate cancer, its effects on the risk for other types of cancer is unclear. In response to the need to confirm these findings, several large placebo-controlled trials designed to further investigate the role of selenium supplementation in prostate cancer prevention are presently under way [100, 101].

Possible Mechanisms. Several mechanisms have been proposed for the cancer prevention effects of selenium, (i) maximizing the activity of selenoenzymes and improving antioxidant status, (ii) improving immune system function, (iii) affecting the metabolism of carcinogens, and (iv) increasing the levels of selenium metabolites that inhibit tumor cell growth. A two-stage mode has been proposed to explain the different anticarcinogenic activities of selenium at different doses. At nutritional or physiologic doses (~40–100 μg/day in adults) selenium maximizes antioxidant selenoenzyme activity and probably enhances immune system function and carcinogen metabolism. At supra-nutritional or pharmacologic levels (~200–300 μg/day in adults) the formation of selenium metabolites, especially methylated forms of selenium, may also exert anticarcinogenic effects.

Immune Function

Selenium deficiency has been associated with impaired function of the immune system. Moreover, selenium supplementation in individuals who are not selenium-deficient appears to stimulate the immune response. In two small studies, healthy [102, 103] and immunosuppressed individuals [104] supplemented with 200 μg/day of selenium as sodium selenite for 8 weeks showed an enhanced immune cell response to foreign antigens compared with those taking a placebo. A considerable amount of basic research also indicates that selenium plays a role in regulating the expression of cytokines [105].

Viral Infection

Selenium deficiency appears to enhance the virulence or progression of some viral infections. The increased oxidative stress resulting from selenium deficiency may induce mutations or changes in the expression of some viral genes. Selenium deficiency results in decreased activity of GPx. Coxsackie virus has been isolated from the blood of some sufferers of Keshan disease, suggesting that it may be a cofactor in the development of this cardiomyopathy associated with selenium deficiency in humans [106].

Safety

Although selenium is required for health, high doses can be toxic. Acute and fatal toxicities have occurred with accidental or suicidal ingestion of gram quantities of selenium. Clinically significant selenium toxicity was reported in 13 individuals taking supplements that contained 27.3 mg per tablet due to a manufacturing error. Chronic selenium toxicity, selenosis, may occur with smaller doses of selenium over long periods of time. The most frequently reported symptoms of selenosis are hair and nail brittleness and loss. Other symptoms may include gastrointestinal disturbances, skin rashes, a garlic breath odor, fatigue, irritability, and nervous system abnormalities. In an area of China with a high prevalence of selenosis, toxic effects occurred with increasing frequency when blood selenium concentrations reached a level corresponding to an intake of 850 μg/day in adults based on the prevention of hair and nail brittleness and loss and early signs of chronic selenium toxicity [81]. The UL of 400 μg/day for adults (see table 10) includes selenium obtained from food which averages about 100 μg/day for adults in the USA as well as selenium from supplements.

Zinc

Zinc is an essential trace element for all forms of life. Clinical zinc deficiency in humans was first described in 1961, when the consumption of diets

Table 10. Tolerable UL for selenium in the USA

Age group	UL, μg/day
Infants 0–6 months	45
Infants 6–12 months	60
Children 1–3 years	90
Children 4–8 years	150
Children 9–13 years	280
Adolescents 14–18 years	400
Adults 19 years and older	400

with low zinc bioavailability (due to high phytic acid) content was associated with 'adolescent nutritional dwarfism' in the Middle East [107]. Since then, zinc insufficiency has been recognized by a number of experts as an important public health issue, especially in developing countries [108].

Food Sources

Shellfish, beef, and other red meats are rich sources of zinc. Nuts and legumes are good plant sources. Zinc bioavailability is relatively high in meat, eggs, and seafood because of the relative absence of compounds that inhibit zinc absorption. The zinc in whole grain products and plant proteins is less bioavailable due to their relatively high content of phytic acid, a compound that inhibits zinc absorption [109]. The enzymatic action of yeast reduces the level of phytic acid in foods. Recently, national dietary surveys in the USA estimated that the average dietary zinc intake was 9 mg/day for adult women and 13 mg/day for adult men [110].

Function

Numerous aspects of cellular metabolism are zinc-dependent. Zinc plays important roles in growth and development, the immune response, neurological function, and reproduction. On the cellular level, the function of zinc can be divided into three categories: (i) catalytic, (ii) structural, and (iii) regulatory [111].

(i) Nearly a hundred different enzymes depend on zinc for their ability to catalyze vital chemical reactions. Zinc-dependent enzymes can be found in all known classes of enzymes [110].

(ii) Zinc plays an important role in the structure of proteins and cell membranes. A finger-like structure, known as a zinc finger motif, stabilizes the structure of a number of proteins. For example, copper provides the catalytic activity for the antioxidant enzyme copper-zinc superoxide dismutase (CuZnSOD) while zinc plays a critical structural role [109, 110]. The structure and function of cell membranes are also affected by zinc. Loss of zinc from biological membranes increases their susceptibility to oxidative damage and impairs their function [112].

(iii) Zinc finger proteins have been found to regulate gene expression by acting as transcription factors. Zinc also plays a role in cell signaling and has been found to influence hormone release and nerve impulse transmission. Recently, zinc has been found to play a role in apoptosis [113].

Nutrient-Nutrient Interactions

Copper. Taking large quantities of zinc ($\geq$50 mg/day) over a period of weeks can interfere with copper bioavailability. High intake of zinc induces the intestinal synthesis of metallothionein. It traps copper within intestinal cells and prevents its systemic absorption. More typical intakes of zinc do not affect copper absorption and high copper intakes do not affect zinc absorption [109].

Iron. Supplemental (38–65 mg/day of elemental iron) but not dietary levels of iron may decrease zinc absorption. This interaction is important in the management of iron supplementation during pregnancy and lactation and has led to the recommendation of zinc supplementation for pregnant and lactating women taking >60 mg/day of elemental iron [114, 115].

Calcium. High levels of dietary calcium impair zinc absorption in animals, but it is uncertain whether this occurs in humans. Increasing the calcium intake by 890 mg/day in the form of milk or calcium phosphate (total calcium intake 1,360 mg/day) reduced zinc absorption and zinc balance in postmenopausal women [116], but increasing the calcium intake of adolescent girls by 1,000 mg/day in the form of calcium citrate malate (total calcium intake 1,667 mg/day) did not affect zinc absorption or balance [117]. Calcium in combination with phytic acid reduces zinc absorption.

Folic Acid. The bioavailability of dietary folate is increased by the action of zinc-dependent enzyme, suggesting a possible interaction between zinc and folic acid. In the past, some studies found low zinc intake to decrease folate absorption while other studies described folic acid supplementation to impair zinc utilization in individuals with marginal zinc status [109, 110]. However, a more recent study found that supplementation with a high dose of folic acid (800 µg/day) for 25 days did not alter zinc status in a group of students being fed with low zinc diets (3.5 mg/day) nor did zinc intake impair folate utilization [118].

Deficiency

Information about zinc deficiency was mostly derived from the study of individuals born with acrodermatitis enteropathica, a genetic disorder resulting from the impaired uptake and transport of zinc. The symptoms of severe zinc

Table 11. RDA for zinc in the USA

Life stage	Age	Males mg/day	Females mg/day
Infants	0–6 months	2	2
Infants	7–12 months	3	3
Children	1–3 years	3	3
Children	4–8 years	5	5
Children	9–13 years	8	8
Adolescents	14–18 years	11	9
Adults	19 years and older	11	9
Pregnancy	18 years and younger	–	12
Pregnancy	19 years and older	–	11
Breast-feeding	18 years and younger	–	13
Breast-feeding	19 years and older	–	12

deficiency include the slowing or cessation of growth and development, delayed sexual maturation, characteristic skin rashes, chronic and severe diarrhea, immune system deficiencies, impaired wound healing, diminished appetite, impaired taste sensation, night blindness, swelling and clouding of the corneas, and behavioral disturbances. Before the cause of acrodermatitis enteropathica was known, patients typically died in infancy. Oral zinc therapy results in the complete remission of symptoms, though it must be maintained indefinitely in individuals with the genetic disorder [119]. Although dietary zinc deficiency is unlikely to cause severe zinc deficiency in individuals without a genetic disorder, zinc malabsorption or conditions of increased zinc loss, such as severe burns or prolonged diarrhea, may also result in severe zinc deficiency.

More recently, is has become apparent that milder zinc deficiency contributes to a number of health problems, especially common in children who live in developing countries. The lack of a sensitive indicator of mild zinc deficiency hinders the scientific study of its health implications. However, controlled trials of moderate zinc supplementation have demonstrated that mild zinc deficiency contributes to impaired physical and neuropsychological development, and increased susceptibility to life-threatening infections in young children [119].

The RDA for zinc is listed for all age groups because infants, children, and pregnant and lactating women are at increased risk of zinc deficiency. Since a sensitive indicator of zinc nutritional status is not readily available, the RDA for zinc was based on a number of different indicators of zinc nutritional status and represents the daily intake likely to prevent deficiency in nearly all individuals in a specific age and gender group [110] (table 11).

Disease Prevention

Growth and Development

In the 1970s and 1980s, several randomized placebo-controlled studies of zinc supplementation in young children with significant growth delays were conducted in Denver, Colorado. Modest zinc supplementation (5.7 mg/day) resulted in increased growth rates compared to placebo [120]. More recently, a number of larger studies in developing countries observed similar results with modest zinc supplementation. A meta-analysis of growth data from zinc intervention trials recently confirmed the widespread occurrence of growth-limiting zinc deficiency in young children, especially in developing countries [121]. Although the exact mechanism for the growth-limiting effects of zinc deficiency are not known, recent research indicates that zinc availability affects cell-signaling systems that coordinate the response to growth-regulating hormone, insulin-like growth factor-1 (IGF-1) [122].

Low maternal zinc nutritional status has been associated with diminished attention in the newborn infant and poorer motor function at 6 months of age. Zinc supplementation has been associated with improved motor development in very low-birth-weight infants, more vigorous activity in Indian infants and toddlers, and more function activity in Guatemalan infants and toddlers [123]. Additionally, zinc supplementation was associated with better neuropsychologic functioning (e.g., attention) in Chinese students, but only when zinc was provided with other micronutrients [124]. Two other studies failed to find an association between zinc supplementation and measures of attention in children diagnosed with growth retardation. Although initial studies suggest that zinc deficiency may depress cognitive development in young children, more controlled research is required to determine the nature of the effect and whether zinc supplementation is beneficial [125].

Immune Function

Adequate zinc intake is essential in maintaining the integrity of the immune system [126] and zinc-deficient individuals are known to experience increased susceptibility to a variety of infectious agents [127]. Age-related declines in immune function are similar to those associated with zinc deficiency, and certain aspects of immune function in the elderly have been found to improve with zinc supplementation [128]. For instance, a randomized placebo-controlled study in men and women over 65 years of age found that a zinc supplement of 25 mg/day for 3 months increased levels of CD4 T cells and cytotoxic T lymphocytes compared to placebo [129]. However, other studies have not found zinc supplementation to improve parameters of immune function, indicating that more research is required before any recommendations regarding zinc and immune system response in the elderly can be made.

Pregnancy Complications

It has been estimated that 82% of pregnant women worldwide are likely to have inadequate zinc intakes. Poor maternal zinc status has been associated with a number of adverse outcomes of pregnancy, including low birth weight, premature delivery, and labor and delivery complications. However, the results of maternal zinc supplementation trials in the USA and developing countries have been mixed [123]. Although some studies have found maternal zinc supplementation to increase birth weight and decrease the likelihood of premature delivery, two recent studies in Peruvian and Bangladeshi women found no difference between zinc supplementation and placebo in the incidence of low birth weight or premature delivery [130, 131]. Supplementation studies designed to examine the effect of zinc supplementation on labor and delivery complications have also generated mixed results, though few have been conducted in zinc-deficient populations [123].

Age-Related Macular Degeneration

A leading cause of blindness in people over the age of 65 is a degenerative disease of the macula known as age-related macular degeneration (AMD). Observational studies have not demonstrated clear associations between dietary zinc intake and the incidence of AMD [132–134]. A randomized controlled trial attracted the interest as it was found that 200 mg/day of zinc sulfate (91 mg/day of elemental zinc) over 2 years reduced the loss of vision in patients with AMD [135]. However, a later trial using the same dose and duration found no beneficial effect in patients with a more advance form of AMD in one eye [136]. A large randomized controlled trial of daily antioxidant (500 mg of vitamin C, 400 IU of vitamin E, and 15 mg of β-carotene) and high-dose zinc (80 mg of zinc and 2 mg of copper) supplementation found that the antioxidant combination plus high-dose zinc and high-dose zinc alone significantly reduced the risk of advanced macular degeneration compared to placebo in individuals with signs of moderate to severe macular degeneration in at least one eye [137]. At present, there is little evidence that zinc supplementation would be beneficial to people with

Table 12. Tolerable UL for zinc in the USA

Age group	UL, mg/day
Infants 0–6 months	4
Infants 7–12 months	5
Children 1–3 years	7
Children 4–8 years	12
Children 9–13 years	23
Adolescents 14–18 years	34
Adults 19 years and older	40

early signs of macular degeneration but further randomized controlled trials are warranted [138].

Safety

Isolated outbreaks of acute zinc toxicity have occurred as a result of consumption of food or beverages contaminated with zinc released from galvanized containers. Signs of acute zinc toxicity are abdominal pain, diarrhea, nausea, and vomiting. Single doses of 225–450 mg of zinc usually induce vomiting. Milder gastrointestinal distress has been reported at doses of 50–150 mg/day of supplemented zinc. Metal fume fever has been reported after the inhalation of zinc oxide fumes. Profuse sweating, weakness, and rapid breathing may develop within 8 h of zinc oxide inhalation and persist 12–24 h after exposure is terminated [109, 110].

The major consequence of long-term consumption of excessive zinc is copper deficiency. Total zinc intakes of 60 mg/day (50 mg supplemental and 10 mg dietary zinc) have been found to result in sign of copper deficiency. In order to prevent copper deficiency, the US UL of intake was set for adults at 40 mg/day, including dietary and supplemental zinc [110] (table 12).

Carotenoids: α-Carotene, β-Carotene, β-Cryptoxanthin, Lycopene, Lutein and Zeaxanthin

Carotenoids are a class of more than 600 naturally occurring pigments synthesized by plants, algae, and photosynthetic bacteria. These colored molecules are the sources of the yellow, orange, and red colors of many plants [139]. Fruits and vegetables provide most of the carotenoids in the human diet; α- and β-carotene, β-cryptoxanthin, lutein, zeaxanthin, and lycopene are the most common carotenoids in the diet.

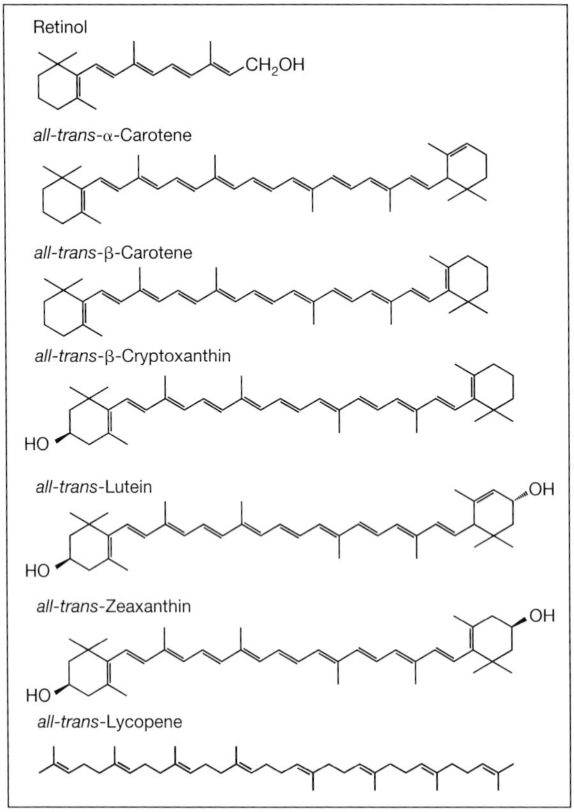

Retinol

CH_2OH

all-trans-α-Carotene

all-trans-β-Carotene

all-trans-β-Cryptoxanthin

HO

all-trans-Lutein

OH

HO

all-trans-Zeaxanthin

OH

HO

all-trans-Lycopene

Food Sources

Carotenoids in foods are mainly in the *all-trans* form although cooking may result in the formation of other isomers. Average intakes for the major carotenoids in the US diet are [140]:

- α-Carotene: adult men, 0.4–0.6 mg/day; adult women, 0.2–0.6 mg/day
- β-Carotene: adult men, 2.1–2.7 mg/day; adult women, 1.6–2.6 mg/day
- Lutein + zeaxanthin: adult men, 2.0–2.3 mg/day; adult women, 1.7–2.0 mg/day
- Lycopene: adult men, 6.6–12.6 mg/day; adult women, 4.3–7.4 mg/day.

α- and β-Carotene are provitamin A carotenoids, i.e. they can be converted by the body to vitamin A. The vitamin A activity of β-carotene in foods is 1/12 that of retinol. Thus, it would take 12 μg of β-carotene from foods to provide the equivalent of 1 μg of retinol. The vitamin A activity of α-carotene from foods is 1/24 that of retinol, so it would take 24 μg of α-carotene from foods to provide the equivalent of 1 μg of retinol. Orange and yellow vegetables like carrots and winter squash are rich sources of α- and β-carotene. Spinach is also a rich source of β-carotene.

Like α- and β-carotene, β-cryptoxanthin is a provitamin A carotenoid. The vitamin A activity of β-cryptoxanthin from foods is 1/24 that of retinol. Orange and red fruits as well as vegetables like sweet red peppers and oranges are particularly rich sources of β-cryptoxanthin.

Lycopene gives tomatoes, pink grapefruit, watermelon, and guava their red color. It has been estimated that 80% of the lycopene in the diet comes from tomatoes and tomato products such as tomato sauce, tomato paste, and ketchup [141]. Lycopene is not a provitamin A carotenoid, i.e. the body cannot convert lycopene to vitamin A.

Lutein and zeaxanthin are both from the class of carotenoids known as xanthophylls. They are not provitamin A carotenoids. Some methods used to quantify lutein and zeaxanthin in foods do not separate the two compounds, so they are mostly reported as lutein + zeaxanthin. Lutein and zeaxanthin are present in a variety of fruits and vegetables. Dark green leafy vegetables, like spinach and kale, are particularly rich sources of lutein and zeaxanthin.

Function

Vitamin A Activity. α-Carotene, β-carotene, and β-cryptoxanthin are provitamin A carotenoids, their essential function recognized in humans is to serve as a source of vitamin A [142].

Antioxidant Activity. In plants, carotenoids have the important antioxidant function of quenching singlet oxygen [143]. Among them, lycopene is one of the most effective quenchers of singlet oxygen [144]. Although important for plants, the relevance of singlet oxygen quenching to human health is less clear. Carotenoids can also inhibit lipid peroxidation, but their actions in humans appear to be more complex [145]. At present, it is unclear whether the biological effects of carotenoids in humans are a result of their antioxidant activity or other non-antioxidant mechanisms.

Light Filtering. The long system of alternating double and single bonds common to all carotenoids allows them to absorb light in the visible range of the spectrum. This feature has particular relevance to the eye, where lutein and zeaxanthin efficiently absorb blue light. Reducing the amount of blue light that reaches the structures of the eye that are critical to vision may protect them from light-induced oxidative damage [146].

Intercellular Communication. Carotenoids can facilitate communication between neighboring cells grown in culture by stimulating the synthesis of connexion proteins [147]. Carotenoids increase the expression of the gene encoding a connexion protein, an effect that appears unrelated to the vitamin A or antioxidant activities of various carotenoids [148].

Immune System Activity. As vitamin A is essential for normal immune system function, it is difficult to determine whether the effects of provitamin A carotenoids are related to their vitamin A activity or other activities of carotenoids. Although some clinical trials have found that β-carotene supplementation improves several biomarkers of immune function [149–151], increasing intakes of lycopene and lutein did not result in similar improvements in immune function biomarkers [152].

Deficiency

Although consumption of provitamin A carotenoids can prevent vitamin A deficiency, no deficiency symptoms have been identified in people consuming low-carotenoid diets if they consume adequate vitamin A [142]. After reviewing the published scientific research, the Food and Nutrition Board of the Institute of Medicine concluded that the existing evidence in 2000 was insufficient to establish a RDA or adequate intake for carotenoids.

Disease Prevention

Cardiovascular Diseases

Evidence that LDL oxidation plays a role in the development of atherosclerosis led to investigations of the role of antioxidant compounds such as carotenoids in the prevention of cardiovascular diseases [153]. A number of case-control and cross-sectional studies have found higher blood levels of carotenoids to be associated with significantly lower measures of carotid artery-intima-media thickness [154–159]. Higher plasma carotenoids at baseline have been associated with significant reductions in cardiovascular disease risk in some prospective studies [160–162], but not in others [163–165]. While the results of several prospective studies indicate that people with higher intakes of carotenoid-rich fruits and vegetables are at lower risk of cardiovascular disease [166–168], it is not yet clear whether this effect is a result of carotenoids or other factors associated with diets high in carotenoid-rich fruits and vegetables.

In contrast to the results of epidemiologic studies suggesting that high dietary intakes of carotenoid-rich fruits and vegetables may decrease cardiovascular disease risk, four randomized controlled studies found no evidence that β-carotene supplements in doses ranging from 20 to 50 mg/day were effective in preventing cardiovascular diseases [169–172]. Based on these results, it has been concluded that there was good evidence that β-carotene supplements provided to benefit in the prevention of cardiovascular disease in middle-aged and older adults [173, 174]. Although diets rich in β-carotene have generally been associated with reduced cardiovascular disease risk in observational studies, there is no evidence that β-carotene supplementation reduces cardiovascular disease risk.

Age-Related Macular Degeneration

In Western countries, degeneration of the macula is the leading cause of blindness in older adults. The only carotenoids found in the retina are lutein and zeaxanthin. By preventing a substantial amount of the blue light entering the eye from reaching underlying structures involved in vision, lutein and zeaxanthin may protect them from light-induced oxidative damage which is thought to play a role in the pathology of AMD [146]. It is also possible, though not proven, that lutein and zeaxanthin act directly to neutralize oxidants formed in the retina. Epidemiologic studies provide some evidence that higher intakes of lutein and zeaxanthin are associated with lower risk of AMD [175]. However, the relationship is not clear. While cross-sectional and retrospective case-control studies found that higher levels of lutein and zeaxanthin in the diet [176–182], blood, and retina were associated with lower incidence of AMD, two prospective cohort studies found no relationship between baseline dietary intakes or serum levels of lutein and zeaxanthin and the risk of developing AMD over time [183–185]. Thus, it seems to be premature to recommend supplements without data from randomized controlled trials [186]. The available scientific evidence suggests that consuming at least 6 mg/day of lutein and zeaxanthin from fruits and vegetables may decrease the risk of AMD [176–178].

The only published randomized controlled trial designed to examine the effect of a carotenoid supplement on the risk AMD used β-carotene in combination with vitamin C, vitamin E, and zinc because lutein and zeaxanthin were not commercially available as supplements at the time the trial began [187]. Although the combination of antioxidants and zinc lowered the risk of developing advanced macular degeneration in individuals with signs of moderate to severe macular degeneration in at least one eye, it is unlikely that the benefit was related to β-carotene since it is not present in the retina. Supplementation of male smokers in Finland with 20 mg/day of β-carotene for 6 years did not decrease the risk of AMD compared to placebo [188].

Cataracts

Ultraviolet light and oxidants can damage proteins in the eye's lens causing structural changes that result in the formation of opacities known as cataracts. As people grow old, cumulative damage to lens proteins often results in cataracts that are large enough to interfere with vision [143].

The potential for increased intakes of lutein and zeaxanthin to prevent or slow the progression of cataracts has been pointed out [146]. Three large prospective cohort studies found that men and women with the highest intakes of foods rich in lutein and zeaxanthin, i.e. spinach, kale, and broccoli, were 20–50% less likely to require cataract extraction [189, 190] or develop cataracts [191]. Additional research is required to determine whether these findings are related specifically to lutein and zeaxanthin intake or to other factors associated with diets high in lutein-rich foods.

In addition it has to be mentioned that β-carotene supplementation (20 mg/day) for more than 6 years did not affect the prevalence of cataracts or the frequency of cataract surgery in male smokers living in Finland [188]. In contrast, a 12-year study of male physicians in the USA found that β-carotene supplementation (50 mg every other day) decreased the risk of cataracts in smokers but not in non-smokers [192]. Two randomized controlled trials examined the effect of an antioxidant combination that included β-carotene, vitamin C, and vitamin E on the progression of cataracts. While one study found no benefit after more than 6 years of supplementation [193], the other study found a small decrease in the progression of cataracts after 3 years of supplementation [194]. Altogether, the results of randomized controlled trials suggest that the benefit of β-carotene in slowing the progression of age-related cataracts does not outweigh the potential risks.

Cancer

Lung Cancer. The results of early observational studies suggested that an inverse relationship between lung cancer risk and β-carotene intake was often assessed by measuring blood levels of β-carotene [195, 196]. More recently, the development of databases for other carotenoids in food has allowed to estimate dietary intakes of total and individual dietary carotenoids more accurately. In contrast to early retrospective studies, recent prospective cohort studies have not consistently found inverse associations between β-carotene intake and lung cancer risk. Analysis of dietary carotenoid intake and lung cancer risk in two large prospective cohort studies in the USA that followed more than 120,000 men and women for at least 10 years revealed no significant association between dietary β-carotene intake and lung cancer risk [197]. However, men and women with the highest intakes of total carotenoids, β-carotene, and lycopene were at significantly lower risk of developing lung cancer than those with the lowest intakes. Dietary intakes of total carotenoids, lycopene, β-cryptoxanthin, lutein, and zeaxanthin, but not β-carotene, were associated with significant reductions in lung cancer in a 14-year study of more than 27,000 Finnish male smokers [198], whereas only dietary intakes of β-cryptoxanthin and lutein, and zeaxanthin were inversely associated with lung cancer risk in a 6-year study of more than 58,000 Dutch men [199]. A recent analysis of the pooled results of six prospective cohort studies in North America and Europe also found no relationship between dietary β-carotene intake and lung cancer risk, although those with the highest β-cryptoxanthin intakes had a risk of lung cancer that was 24% lower than those with the lowest intakes [200]. While smoking remains the strongest risk factor for lung cancer, results of recent prospective studies using accurate estimates of dietary carotenoid content suggest that diets rich in a number of carotenoids – not only β-carotene – may be associated with reduced lung cancer risk.

In addition, the effect of β-carotene supplementation on the risk of developing lung cancer has been examined in three large randomized placebo-controlled trials. In Finland, the α-Tocopherol β-Carotene (ATBC) cancer prevention trial evaluated the effects of 20 mg/day of β-carotene and/or 50 mg/day of α-tocopherol on more than 29,000 male smokers [169], and in the USA, the β-Carotene and Retinol Efficacy Trial (CARET) evaluated the effects of a combination of 30 mg/day of β-carotene and 25,000 IU/day of retinol (vitamin A) in more than 18,000 men and women who were smokers, former smokers, or had a history of occupational asbestos exposure [201]. Unexpectedly, the risk of lung cancer in the groups taking β-carotene supplements was increased by 16% after 6 years in the ATBC participants and increased by 28% after 4 years in the CARET participants. The Physicians Health Study (PHS) examined the effect of β-carotene supplementation (50 mg every other day) on cancer risk in more than 22,000 male physicians in the USA, of whom only 11% were current smokers [170]. In that lower risk population, β-carotene supplementation for more than 12 years was not associated with an increased risk of lung cancer. Although the reasons for the increase in lung cancer risk are not yet clear, experts feel that the risks of high-dose β-carotene supplementation outweigh any potential benefits for cancer prevention, especially in smokers or other high-risk populations [173, 202].

Prostate Cancer. The results of several prospective cohort studies suggest that lycopene-rich diets are associated with significant reduction in the risk of prostate cancer [203]. In a study of more than 47,000 health professionals followed for 8 years, those with the highest lycopene intake had a risk of prostate cancer that was 21% lower than those with the lowest lycopene intake [204]. Those with the highest intakes of tomatoes and tomato products (accounting for 82% of total lycopene intake) had a risk of prostate cancer that was 35% lower and a risk of aggressive prostate cancer that was 53% lower than those with the lowest intakes. Similarly, a prospective study of Seventh Day Adventist men found that those who reported the highest tomato intakes were at significantly lower risk of prostate cancer [205] and that those with the highest plasma lycopene levels were at significantly lower risk of developing aggressive prostate cancer [206]. However, dietary lycopene intake was not related to prostate cancer risk in a prospective study of more than 58,000 Dutch men [207]. Thus, it is not clear whether the prostate cancer risk reduction observed in some epidemiologic studies is related to lycopene itself, other compounds in tomatoes, or other factors associated with lycopene-rich diets.

Bioavailability of Carotenoids

The bioavailability of carotenoids is influenced by a number of factors. In general, purified carotenoids in oil (supplements) are more bioavailable than carotenoids in foods [208]. In particular, the bioavailability of β-carotene from supplements is much higher than from foods. One study found that the bioavailability of β-carotene from spinach was only 14% of that of purified β-carotene in oil [209]. In contrast, the bioavailability of lutein from spinach was 67% of that of purified lutein in oil.

The relatively low bioavailability of carotenoids from foods is partly due to the fact that they are associated with proteins in the plant matrix. Carotenoids are associated with chloroplasts in green leafy vegetables and chromoplasts in fruit. Chopping, homogenizing, and cooking disrupt the plant matrix, increasing the bioavailability of carotenoids [208]. The bioavailability of lycopene from tomatoes is substantially improved by heating tomatoes in oil [210, 211].

Safety

High doses of β-carotene supplements ($\geq$30 mg/day) and the consumption of large amounts of carotene-rich foods have resulted in carotenodermia and, analogously, high intakes of lycopene-rich foods or supplements may result in lyopenodermia. Adverse effects of lutein and zeaxanthin have not been reported.

Unlike vitamin A, high doses of β-carotene taken by pregnant women have not been associated with increased risk of birth defects. However, the safety of high-dose β-carotene supplements in pregnancy and lactation has not been well studied. Although there is no reason to limit dietary β-carotene intake, pregnant and breast-feeding women should avoid consuming >3 mg/day (5,000 IU/day) of β-carotene from supplements unless they are prescribed under medical supervision [212].

Interactions among Drugs and Carotenoids

The cholesterol-lowering agents colestyramine and colestipol can reduce absorption of fat-soluble vitamins and carotenoids as can mineral oil and orlistat, a drug used to treat obesity. Colchicine, a drug used to treat gout, can cause intestinal malabsorption. However, long-term use of 1–2 mg/day of colchicine did not affect serum β-carotene levels [213]. Increasing gastric pH through the use of proton pump inhibitors decreased the absorption of a single dose of a β-carotene supplement but it is not known if the absorption of dietary carotenoids is affected [214–216].

The results of metabolic studies suggest that high doses of β-carotene compete with lutein and lycopene for absorption when consumed at the same time. However, the consumption of high-dose β-carotene supplements did not adversely affect serum carotenoid levels in long-term clinical trials [217–220].

Polyphenols and Flavonoids

Phenylchromane derivatives having a 2-phenylchromane (= flavane) skeleton are called flavonoids. They represent the single, most widely occurring group of phenolic phytochemicals. Flavonoids are classified according their oxidation level of their central C ring. These variations define the families of flavones, flavanols, flavanones, flavonols, flavandiols (leukoanthocyanidins) and flavylium salts (anthocyanidins). Structural differences between the various members of these families are mainly caused by the hydroxylations in different positions, methylations of individual hydroxyl groups, and glycosidation by various sugars. The most common sugars are *D*-glucose, *L*-rhamnose, *D*-galactose, *D*-glucuronic acid, *D*-galacturonic acid, *L*-arabinose, and *D*-xylose. β-Glycosidation prevails in the *D*-series, while *L*-sugars are α-glycosidically bound. In total, several thousand different flavonoids have been described [221].

Food Sources
In the human diet, chlorogenic acid (e.g. from coffee, carrots), ferulic acid (e.g. from cereals), flavonols (e.g. from onions, vegetables, tea, apples), catechin and other flavan-3-ols (e.g. from apples, grapes, chocolate), isoflavones

(e.g. from soybeans) as well as lignans (e.g. from cereals) constitute the major classes.

The lack of reliable compositional data and the fact that intake of any single plant phenol will be highly dependent on the types of food plants consumed, means that it is not possible to provide definitive values of intake in human populations. In the 1970s, it was generally assumed that the average intake of dietary flavonoids is in the range of 1 g/day. This figure has been questioned later. Related studies of the flavonoid content of common human beverages indicate consumption rates in the lower milligram ranges, in which tea and onions are major contributors.

Function

The presence of a phenolic group in a natural flavonoid would be expected to provide antimicrobial activity and the addition of further phenolic groups might be expected to enhance this activity. In fact, one of the undisputed functions of flavonoids and related polyphenols is their role in protecting plants against microbial invasion. This not only involves their presence in plants as constitutive agents but also their accumulation as phytoalexins in response to microbial attack [222].

Yet one further property of flavonoids that has been researched recently has had antiviral activity, most notably against the human immune deficiency virus (HIV). Some flavonoids appear to have direct inhibitory activity on the virus. That is apparently true for baicalin (5,6,7-trihydroxyflavone-7-glucoronide) from *Scutellaria baicalensis* [223]. Other flavonoids are inhibitory to enzymes required for viral replication.

In addition, it is now generally accepted that flavonoids, along with other plant polyphenols, play a role in protecting plants against both insect and herbivorous mammals, i.e. they are active in plant-animal interactions. In recent years, attention has mainly been focused on simple phenolic constituents or on the polymeric proanthocyanidins, but further research has also been concerned with low-molecular-weight flavones, flavonols, and isoflavones [224].

Disease Prevention

In the last decade, the medicinal properties of flavonoids came increasingly into the center of research. The most important effects will briefly be discussed in the following.

Antioxidant Properties and Enzyme Inhibition

Flavonoids have been shown to act as scavengers of various oxidizing species, such as superoxide anion, hydroxyl radical or singlet oxygen. Flavonoids do not react specifically with a single species and so a number of different evaluation methods have been developed which makes comparison of the various studies very difficult. The structural conditions for the antioxidation activity have been reviewed recently [221].

A possible contributory mechanism to the antioxidant activity of flavonoids is their ability to stabilize membranes by decreasing membrane fluidity. A series of representative flavonoids partition into the hydrophobic core of the membrane, causing a dramatic decrease in lipid fluidity in this region of the membrane [225].

Flavonoids are known to inhibit key enzymes in mitochondrial respiration [226]. Some flavonoids also inhibit the enzyme xanthine oxidase which catalyzes the oxidation of xanthine and hypoxanthine to uric acid. During the re-oxidation of xanthine oxidase, both superoxide radicals and hydrogen peroxide are produced. Obviously, flavons show higher inhibitory effect than flavonols and hydroxyl groups at both C-3 and C-3' are essential for high superoxide scavenging activity [227].

The in vitro antioxidative activities have been recognized for decades, but it is still not clear whether there are in vivo beneficial effects.

Anti-Inflammatory Activity

Flavonoids may inhibit the cyclooxygenase and/or the 5-lipoxygenase pathways of arachidonate metabolism. The various structural-related effects have been reviewed comprehensively [221].

Vascular Activity

Flavonoids may act in a number of different ways on the various components of blood, such as platelets, monocytes, LDL, and smooth muscles. Platelets are key participants in atherogenesis and proinflammatory indicators, such as thromboxane A_2, PAF and serotonin are produced from them. Flavonoids may inhibit platelet adhesion, aggregation and secretion [221].

In a survey of 65 flavonoids for procoagulant activity, 18 were found to inhibit the interleukin-1-induced expression of tissue factor on human monocytes [228]. Flavonols, such as kaempferol, quercetin and myricetin have been shown to inhibit adenosine deaminase activity in the endothelial cells of the aorta whereas flavones were found to be inactive [229].

Flavonoids have also been shown to be potent inhibitors of the oxidative modification of LDL by macrophages. They also inhibit the cell-free oxidation of LDL mediated by copper sulfate [230].

Coronary Heart Disease

In most countries a high intake of saturated fats is strongly correlated with high mortality from CHD, but this is not the case in some regions of France (so-called 'French paradox'). This anomaly has been attributed to the regular intake of red wine in the diet [231]. Epicatechin and quercetin might be more important than resveratrol in reducing CHD. It is suggested that the combination of antioxidant phenolics in wine may protect against atherogenesis with regular long-term consumption. There are several other studies considering the effect of flavonoids on CHD and the role of dietary antioxidant flavonoids protecting against CHD has been more widely reviewed [232].

Cytotoxic Antitumor Activity

There have been many bioassay-guided searches for cytotoxic antitumor agents in plants, especially those known to be used in folk medicine for this purpose. This has led to the isolation and identification of numerous active constituents from all the different flavonoid classes. However, the choice and number of all lines used in these bioassays has been very variable and it is difficult to draw general conclusions from them. The literature has been reviewed comprehensively [221].

Estrogenic Activity

The main group of flavonoids that is well known to possess estrogenic activities are the isoflavones, such as genistein. In a normal human diet the presence of such active flavonoids is generally considered to be harmless because no single phytoestrogen is present in sufficient quantity to have physiological consequences. However, this may not be the case for vegetarians, especially those who eat a large percentage of legumes in their diet which has a high isoflavonoid content, such as soya and pulses.

Other Biological Activities

It is well known that some flavonoids can act as antispasmolytic agents by relaxing smooth muscles in various parts of the mammalian body. An impressive example for a flavone with anxiolytic and anticonvulsive activity is the most recently studied hispidulin [233]. Flavonoids may also exhibit antibacterial [234], antifungal [235], and antimalaria [236] activities.

Safety

Risk/benefit evaluations are under investigation worldwide, however, the lack of sufficient data does not allow to draw final conclusions. Nonetheless, the recently published statements on safety aspects of functional food including flavonoids show that scientists clearly stress the importance to substantially evaluate the safety to health, functionality and claims as well as recommend observations after the market introduction [237].

α-Lipoic Acid

Also known as thioctic acid, α-lipoic acid is a naturally occurring compound that is synthesized by plants and animals, including humans. α-Lipoic acid contains two sulfur molecules that can be oxidized or reduced. This feature allows α-lipoic acid to function as a cofactor for several important enzymes as well as a potent antioxidant. Only the *R*-isomer of α-lipoic acid is synthesized naturally. Conventional chemical synthesis of α-lipoic acid results in a 50/50 R:S (racemic) mixture of the two optical isomers [238].

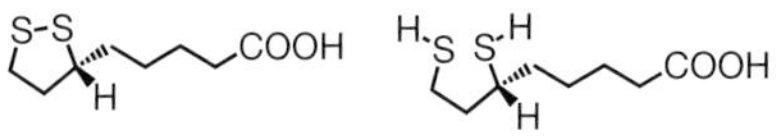

(R) Lipoic acid   (R) Dihydro lipoic acid

Food Sources

Most α-lipoic acid in food is derived from lipoamide-containing enzymes and is bound to lysine (lipoyllysine). Animal tissues that are rich in lipoyllysine include kidney, heart, and liver, whereas lipoyllysine-rich plant sources comprise spinach, broccoli, and tomatoes. Somewhat lower amounts of lipoyllysine have been measured in peas, Brussels sprouts, and rice bran.

Digestive enzymes do not break the bond between α-lipoic acid and lysine effectively. Thus, it has been hypothesized that most dietary α-lipoic acid is absorbed as lipoyllysine, and free α-lipoic acid has not been detected in the circulation of humans who are not taking α-lipoic acid supplements. Although α-lipoic acid is found in a wide variety of foods from plant and animal sources, quantitative information on the α-lipoic acid content of food is limited.

Function

In its protein-bound form, *R*-α-lipoic acid is a required cofactor for several multi-enzyme complexes inside the mitochondria. The pyruvate dehydrogenase complex catalyzes the conversion of pyruvate to acetyl-CoA; the α-ketoglutarate dehydrogenase complex catalyzes the metabolism of three amino acids, leucine, isoleucine, and valine. The glycine cleavage system is a multi-enzyme complex that catalyzes the formation of 5,10-methylene tetrahydrofolate, an important cofactor in nucleic acid synthesis.

When large amounts of free α-lipoic acid are available, e.g., under supplementation, α-lipoic acid is also able to function as an antioxidant [239]. Free α-lipoic acid is rapidly taken up by cells and reduced intracellularly to α-dihydrolipoic acid (DHLA). DHLA is the only form that functions directly as an antioxidant [240]. Because DHLA is rapidly eliminated from cells, the extent to which its antioxidant effects can be sustained remains unclear. DHLA is a potent reducing agent and has the capacity to regenerate a number of oxidized forms, i.e. of vitamin C, glutathione, and coenzyme Q_{10} which are able to regenerate oxidized α-tocopherol, forming an antioxidant network. DHLA can be regenerated from α-lipoic acid through the activity of enzymes present in cells [238].

Certain free metal ions like iron and copper can induce oxidative damage by catalyzing reactions that generate highly reactive free radicals. Both α-lipoic acid and DHLA may chelate or bind metal ions in a way that prevents them from generating free radicals [238]. At present, this property has only been demonstrated in vitro.

Glutathione is an important water-soluble antioxidant that is synthesized from the sulfur-containing amino acid cysteine. The availability of cysteine inside a cell determines its rate of glutathione synthesis. Although increases in intracellular DHLA are short-lived, DHLA may also improve intracellular antioxidant capacity by inducing glutathione synthesis [238].

The protein α_1-antiprotease is an inhibitor of the enzyme elastase. Oxidation inactivates α_1-antiprotease, leading to increased activity of elastase and degradation of elastin in the lungs, a process that has been implicated in chronic obstructive pulmonary disease (COPD). In vitro, DHLA can act as a reducing factor for the enzyme, peptide methionine sulfoxide reductase (PMSR) which can reduce and reactive oxidized α_1-antiprotease [241]. Whether α-lipoic acid contributes to the repair of oxidized proteins in living organisms remains to be determined.

Nuclear factor-κB (NF-κB) is known as a transcription factor, as it is able to bind to DNA and affect the rate of transcription of certain genes that have NF-κB binding sites. NF-κB plays an important role in regulating genes related to inflammation and the pathology of a number of diseases, including atherosclerosis, cancer and diabetes [242]. Physiologically relevant concentrations of α-lipoic acid have been found to inhibit the activation of NF-κB when added to cells in culture [243].

AP-1 is another transcription factor that can be affected by both reactive oxygen species (ROS) and certain antioxidants within cells. Treating cells in culture with DHLA has been found to inhibit the activity of AP-1 by decreasing the expression of the gene for c-fos, one of the proteins that makes up the functional AP-1 complex [244].

Deficiency

α-Lipoic acid deficiency has not been described, suggesting that humans are able to synthesize enough to meet their needs of enzyme cofactors. Increased destruction of the cofactor form of α-lipoic acid may underlie the pathology of some diseases. In arsenic toxicity, arsenic can form a complex with α-lipoic acid in dehydrogenase enzymes, leaving it inactive [239]. Circulating antibodies to lipoamide-containing enzyme subunits have been isolated in patients with primary biliary cirrhosis [245].

Disease Prevention

In aging rats, short-term dietary supplementation with R-α-lipoic acid has been found to decrease mitochondrial ROS production and improve mitochrondrial function [246, 247]. A series of studies in aged rats found that combined dietary supplementation of R-α-lipoic acid and acetyl-L-carnitine improved mitochondrial energy metabolism, decreased oxidative stress, increased physical activity, and improved measures of short-term memory [248, 249]. As these findings are very encouraging, the researchers caution that these studies used relatively high doses of the compounds only for 1 month. It is not yet known whether taking relatively high doses of R-α-lipoic acid and acetyl-L-carnitine will be for the benefit of aging rats in the long term or will have similar effects in humans.

Diabetes mellitus

Pharmacologic doses of α-lipoic acid, i.e. many times higher than the amount a person could synthesize or obtain from foods, have been prescribed to treat diabetic patients in Germany since the late 1960s [250]. Data from animal studies suggest that the R-isomer may be more effective in improving insulin sensitivity than the S-isomer [251, 252], but this possibility has not been tested in any published human trials.

Oxidative Stress. A number of studies in individuals with diabetes (types 1 and 2) indicate that they are under increased oxidative stress, a condition that is believed to contribute to the vascular and neurological complications of diabetes. Although α-lipoic acid supplementation has been found to reduce measures of oxidative stress in animal models of diabetes, evidence that α-lipoic acid reduced oxidative stress in humans with diabetes is limited. In a non-randomized cross-sectional study, 33 patients with type 1 or type 2 diabetes who had taken 600 mg/day of α-lipoic acid orally for at least 3 months, had lower levels of plasma lipid peroxidation than did 74 diabetics who did not take α-lipoic acid [253]. An intervention trial in 10 diabetic patients found that plasma lipid peroxides were significantly lower after taking 600 mg/day of α-lipoic acid orally for 60 days compared to baseline [254]. Oral α-lipoic acid supplementation (600 mg/day) has been found to decrease NF-κB activation in the white blood cells of type 1 diabetics [255] and patients with diabetic nephropathy (kidney damage). The formation of advanced glycation end products (AGEP) also leads to glucose-mediated damage in diabetes. α-Lipoic acid has been found to prevent the formation of AGEP in vitro [256].

Diabetic Peripheral Neuropathy. Over one-third of diabetics develop peripheral neuropathy. In addition to the pain and disability caused by diabetic neuropathy, it is a leading cause of lower limb amputation in diabetics [257]. The results of several large randomized controlled trials indicate that maintaining blood glucose at nearly normal levels is the most important step in decreasing the risk of diabetic neuropathy. However, such intensive blood glucose control may not be achievable in all diabetic patients.

Oxidative stress has been implicated in the pathology of diabetic neuropathy, and α-lipoic acid is approved for the treatment of diabetic neuropathy in Germany. At least 15 clinical trials have examined the effect of α-lipoic acid treatment on symptoms of diabetic neuropathy with mixed results, especially in smaller studies. Modest benefits have been observed in several large multicenter trials. More than 300 type 2 diabetics with symptomatic peripheral neuropathy were randomly assigned to intravenous treatment with 100, 600 or 1,200 mg/day of α-lipoic acid or placebo for 3 weeks [258]. Symptom scores were significantly improved in those that received intravenous infusions of at least 600 mg/day of α-lipoic acid compared to placebo. A subsequent multicenter trial randomly assigned 509 type 2 diabetics with symptomatic peripheral neuropathy to one of different treatments [259]. Although symptom scores did not differ significantly from baseline in any of the groups, assessments of sensory and motor deficits by trained physicians were significantly improved after 3 weeks of intravenous α-lipoic acid therapy and non-significantly improved at the end of 6 months of oral α-lipoic acid therapy. A smaller randomized controlled trial examined the effect of long-term oral α-lipoic acid supplementation on the results of electrophysiological nerve conduction studies in 65 diabetic patients with symptomatic peripheral neuropathy [260]. Those who took α-lipoic acid showed significant improvements in 3 out of 4 nerve conduction assessments compared to those who took placebo.

Overall, the available research suggests that oral doses of at least 600 mg/day of α-lipoic acid may offer some benefit in the alleviation of neuropathy symptoms and deficits, especially when used in conjunction with effective treatment aimed at normalizing blood glucose levels.

Vascular Complications. Endothelial function in individuals with diabetes (types 1 and 2) is often impaired and diabetics are at increased risk for vascular disease. Several small preliminary studies in humans have examined the effect of α-lipoic acid administration on endothelial function. In one study, intra-arterial infusions of α-lipoic acid improved endothelium-dependent vasodilation in 39 diabetic patients, but not in 11 healthy controls [261]. Oral supplementation of 1,200 mg/day of α-lipoic acid for 6 weeks improved a measure of capillary perfusion in the fingers of 8 diabetic patients with peripheral neuropathy [262]. In an uncontrolled, non-randomized study of 84 diabetic

patients, plasma thrombomodulin levels, a marker of compromised endothelial function, decreased significantly in the 35 diabetics that took 600 mg/day of α-lipoic acid orally over 18 months, whereas thrombomodulin levels increased significantly in those who did not take α-lipoic acid over the same period [263]. As the results of these small, uncontrolled trials are encouraging, long-term placebo-controlled studies are needed before it can be determined whether α-lipoic acid supplementation can reduce the risk of vascular complications in individuals with diabetes.

Safety

There is evidence that the enantiomers of α-lipoic acid have different biological activities. Within the mitochondria, *R*-α-lipoic acid is reduced to DHLA, the more potent antioxidant, 28 times faster than *S*-α-lipoic acid. However, in the cytosol *S*-α-lipoic acid is reduced to DHLA twice as fast as *R*-α-lipoic acid. One study in humans found *R*-α-lipoic acid to be more bioavailable than *S*-α-lipoic acid when taken orally [264]. *R*-lipoic acid was more effective than *S*-lipoic acid in enhancing insulin-stimulated glucose transport and metabolism insulin-resistant rat skeletal muscle [250], and *R*-α-lipoic acid was more effective than racemic α-lipoic acid and *S*-α-lipoic acid in preventing cataracts in rats [265]. Almost all studies of α-lipoic acid supplementation in humans have been performed using racemic α-lipoic acid. At present, it is not known whether *R*-α-lipoic acid is more effective as an antioxidant than racemic lipoic acid when taken by humans in pharmacologic doses.

In general, α-lipoic acid doses of 600 mg/day have been well tolerated. Doses as high as 1,200 mg/day (600 mg, twice a day) for 2 years and 1,800 mg/day (600 mg, 3 times a day) for 3 weeks did not result in adverse effects when given to patients with diabetic neuropathy under medical supervision. There are no reports of toxicity from α-lipoic acid overdose in humans. In individuals with diabetes and/or impaired glucose tolerance, α-lipoic acid supplementation may lower blood glucose levels. Individuals on diabetic medications should monitor blood glucose levels. Diabetic medication doses may need to be adjusted to avoid hypoglycemia. As controlled safety studies in pregnant and lactating women are not available, the use of α-lipoic acid supplements by pregnant or breast-feeding women is not recommended [266].

Acknowledgement

The support kindly provided by FRUIT, International Fruit Foundation, Heidelberg, is greatly acknowledged.

References

1   Carr AC, Frei B: Toward a new recommended dietary allowance for vitamin C based on antioxidant and health effects in humans. Am J Clin Nutr 1999;69:1086–1107.
2   Simon JA, Hudes ES: Serum ascorbic acid and gallbladder disease prevalence among US adults: The Third National Health and Nutrition Examination Survey (NHANES III). Arch Intern Med 2000;160:931–936.
3   Carr A, Frei B: Does vitamin C act as a pro-oxidant under physiological conditions? Faseb J 1999; 13:1007–1024.
4   Food and Nutrition Board, Institute of Medicine: Vitamin C. Dietary Reference Intakes for Vitamin C, Vitamin E, Selenium, and Carotenoids. Washington, National Academy Press, 2000, pp 95–185.
5   Losonczy KG, Harris TB, Havlik RJ: Vitamin E and vitamin C supplement use and risk of all-cause and coronary heart disease mortality in older persons: The established populations for epidemiologic studies of the elderly. Am J Clin Nutr 1996;64:190–196.
6   Kushi LH, Folsom AR, Prineas RJ, Mink PJ, Wu Y, Bostick RM: Dietary antioxidant vitamins and death from coronary heart disease in postmenopausal women. N Engl J Med 1996;334: 1156–1162.
7   Enstrom JE, Kanim LE, Klein MA: Vitamin A intake and mortality among a sample of the United States population. Epidemiology 1992;3:194–202.
8   Enstrom JE: Counterpoint-vitamin C and mortality. Nutr Today 1993;28:28–32.
9   Osganian SK, Stamper MJ, Rimm E, Spiegelman D, Hu FB, Manson JE, Willett WC: Vitamin C and risk of coronary heart disease in women. J Am Coll Cardiol 2003;42:246–252.
10 Khaw KT, Bingham S, Welch A, Luben R, Wareham N, Oakes S, Day N: Relation between plasma ascorbic acid and mortality in men and women in EPIC-Norfolk prospective study: A prospective population study. European Prospective Investigation into Cancer and Nutrition. Lancet 2001;357: 657–663.
11 Levine M, Wang Y, Padayatty SJ, Morrow J: A new recommended dietary allowance of vitamin C for healthy young women. Proc Natl Acad Sci USA 2001;98:9842–9846.
12 Frei B: To C or not to C, that is the question! J Am Coll Cardiol 2003;42:253–255.
13 Steinmetz KA, Potter JD: Vegetables, fruit, and cancer prevention: A review. J Am Diet Assoc 1996;96:1027–1039.
14 Zhang S, Hunter DJ, Forman MR, Rosher BA, Speizer FE, Colditz GA, Manson DE, Hankinson SE, Willett WC: Dietary carotenoids and vitamins A, C and E and risk of breast cancer. J Natl Cancer Inst 1999;91:547–556.
15 Michels KB, Holmberg L, Begkvist L, Ljung H, Bruce A, Wolk A: Dietary antioxidant vitamins, retinol, and breast cancer incidence in a cohort of Swedish women. Int J Cancer 2001;91:563–567.
16 Feiz HR, Mobarhan S: Does vitamin C intake slow the progression of gastric cancer in *Heliobacter pylori*-infected populations? Nutr Rev 2002;60:34–36.
17 Jacques PF, Chylack LT Jr, Hankinson SE, Khu PM, Rogers G, Fried J, Tung W, Wolfe JK, Padhye N, Willett WC, Taylor A: Long-term nutrient intake and early age-related nuclear lens opacities. Arch Ophthalmol 2001;119:1009–1019.
18 Simon JA, Hudes ES: Serum ascorbic acid and other correlates of self-reported cataract among older Americans. J Clin Epidemiol 1999;52:1207–1211.
19 AREDS Report No 9, 2001 (II): A randomized, placebo-controlled, clinical trial of high-dose supplementation with vitamins C and E and β-carotene for age-related cataract and vision loss. Arch Ophthalmol 2001;119:1439–1452.
20 Traber MG: Utilization of vitamin E. Biofactors 1999;10:115–120.
21 Food and Nutrition Board, Institute of Medicine: Vitamin E: Dietary Reference Intakes for Vitamin C, Vitamin E, Selenium, and Carotenoids. Washington, National Academy Press, 2000, pp 186–283.
22 Traber MG: Does vitamin E decrease heart attack risk? Summary and implications with respect to dietary recommendations. J Nutr 2001;131:395–397.
23 Christen S, Woodall AA, Shigenaga MK, Southwell-Keely PT, Duncan MW, Ames BN: γ-Tocopherol traps mutagenic electrophiles such as NO(X) and complements α-tocopherol: Physiological implications. Proc Natl Acad Sci USA 1997;94:3217–3222.

24   Li D, Saldeen T, Mehta JL: γ-Tocopherol decreases ox-LDL-mediated activation of nuclear factor-κB and apoptosis in human coronary artery endothelial cells. Biochem Biophys Res Commun 1999; 259:157–161.

25   Helzlsouer KJ, Huang HY, Alberg AJ, Hoffman S, Burke A, Norkus EP, Morris JS, Comstock GW: Association between α-tocopherol, γ-tocopherol, selenium, and subsequent prostate cancer. J Natl Cancer Inst 2000;92:2018–2023.

26   Jiang Q, Christen S, Shigenaga MK, Ames BN: γ-Tocopherol, the major form of vitamin E in the US diet, deserves more attention. Am J Clin Nutr 2001;74:714–722.

27   Traber MG: Vitamin E; in Shils M, Olson JA, Shike M, Ross AC (eds): Nutrition in Health and Disease, ed 9. Baltimore, Williams & Wilkins, 1999, pp 347–362.

28   Ford ES, Sowell A: Serum α-tocopherol status in the US population: Findings from the Third National Health and Nutrition Examination Survey. Am J Epidemiol 1999;150:290–300.

29   Knekt P, Reunanen A, Jarvinen R, Seppanen R, Heliovaara M, Aromaa A: Antioxidant vitamin intake and coronary mortality in a longitudinal study. Am J Epidemiol 1994;139:1180–1189.

30   Kushi LH, Folsom AR, Prineas RJ, Mink PJ, Wu Y, Bostick RM: Dietary antioxidant vitamins and death from coronary heart disease in postmenopausal women. N Engl J Med 1996;334: 1156–1162.

31   Rimm EB, Stampfer MJ, Ascherio A, Giovannucci E, Colditz GA, Willett WC: Vitamin E consumption and the risk of coronary heart disease in men. N Engl J Med 1993;328:1450–1456.

32   Stampfer MJ, Hennekens CH, Manson JE, Colditz GA, Rosner B, Willett WC: Vitamin E consumption and the risk of coronary disease in women. N Engl J Med 1993:328:1444–1449.

33   Gale CR, Ashurst HE, Powers HJ, Martyn CN: Antioxidant vitamin status and carotid atherosclerosis in the elderly. Am J Clin Nutr 2001;74:402–408.

34   McQuillan BM, Hung J, Beilby JP, Nidorf M, Thompson PL: Antioxidant vitamins and the risk of carotid atherosclerosis. The Perth Carotid Ultrasound Disease Assessment Study (CUDAS). J Am Coll Cardiol 2001;38:1788–1794.

35   Heinonen OP, Albanes D, Virtamo J, et al: Prostate cancer and supplementation with α-tocopherol and β-carotene: Incidence and mortality in a controlled trial. J Natl Cancer Inst 1998;90:440–446.

36   Jacques PF: The potential preventive effects of vitamins for cataract and age-related macular degeneration. Int J Vitam Nutr Res 1999;69:198–205.

37   Gale CR, Hall NF, Phillips DI, Maryn CN: Plasma antioxidant vitamins and carotenoids and age-related cataract. Ophthalmology 2001;108:1992–1998.

38   AREDS Report No 9, 2001: A randomized, placebo-controlled, clinical trial of high-dose supplementation with vitamins C and E and β-carotene for age-related cataract and vision loss. Arch Ophthalmol 2001;119:1439–1452.

39   Teikari JM, Rautalahti M, Haukka J, Jarvinen P, Hartman AM, Virtamo J, Albauer D, Heinonen D: Incidence of cataract operations in Finnish male smokers unaffected by α-tocopherol or β-carotene supplements. J Epidemiol Commun Health 1998;52:468–472.

40   Meydani SN, Meydani M, Blumberg JB, Leka LS, Siber G, Loscewski R, Thompson C, Pedrosa MC, Diamond RD, Stollar BJ: Vitamin E supplementation and in vivo immune response in healthy elderly subjects. A randomized controlled trial. JAMA 1997;277:1380–1386.

41   Han SN, Meydani SN: Vitamin E and infectious diseases in the aged. Proc Nutr Soc 1999;58: 697–705.

42   Linder MC, Hazegh-Azam M: Copper biochemistry and molecular biology. Am J Clin Nutr 1996; 63:797S–811S.

43   Uauy R, Olivares M, Gonzalez M: Essentiality of copper in humans. Am J Clin Nutr 1998; 67(suppl):952–959.

44   Turnlund JR: Copper; in Shils M, Olson JA, Shike M, Ross AC (eds): Nutrition in Health and Disease, ed 9. Baltimore, Williams & Wilkins, 1999, pp 241–252.

45   Harris ED: Copper; in O'Dell BL, Sunde RA (eds): Handbook of Nutritionally Essential Minerals. New York, Dekker, 1997, pp 231–273.

46   Food and Nutrition Board, Institute of Medicine: Copper. Dietary Reference Intakes for Vitamin A, Vitamin K, Boron, Chromium, Copper, Iodine, Iron, Manganese, Molybdenum, Nickel, Silicon, Vanadium, and Zinc. Washington, National Academy Press, 2001, pp 224–257.

47 Johnson MA, Fischer JG, Kays SE: Is copper an antioxidant nutrient? Crit Rev Food Sci Nutr 1992;32:1–31.

48 Finley EB, Cerklewski FL: Influence of ascorbic acid supplementation on copper status in young adult men. Am J Clin Nutr 1983;37:553–556.

49 Percival SS, Kauwell GP, Bowser E, Wagner M: Altered copper status in adult men with cystic fibrosis. J Am Coll Nutr 1999;18:614–619.

50 Fox PL, Mazumder B, Ehrenwald E, Mukhopadhyay CK: Ceruloplasmin and cardiovascular disease. Free Radic Biol Med 2000;28:1735–1744.

51 Jones AA, DiSilvestro RA, Coleman M, Wagner TL: Copper supplementation of adult men: Effects on blood copper enzyme activities and indicators of cardiovascular disease risk. Metabolism 1997;46:1380–1383.

52 Ford ES: Serum copper concentration and coronary heart disease among US adults. Am J Epidemiol 2000;151:1182–1188.

53 Klevay LM: Cardiovascular disease from copper deficiency – A history. J Nutr 2000;130(suppl): 489–492.

54 Kinsman GD, Howard AN, Stone DL, Mullins PA: Studies in copper status and atherosclerosis. Biochem Soc Trans 1990;18:1186–1188.

55 Mielcarz G, Howard AN, Mielcarz B, Williams NR, Rajput-Williams J, Nigdigar SV, Stone DL: Leucocyte copper, a marker of copper body status is low in coronary artery disease. J Trace Elem Med Biol 2001;15:31–35.

56 Wang XL, Adachi T, Sim AS, Wilcken DE: Plasma extracellular superoxide dismutase levels in an Australian population with coronary artery disease. Arterioscler Thromb Vasc Biol 1998;18: 1915–1921.

57 Klevay LM: Lack of a recommended dietary allowance for copper may be hazardous to your health. J Am Coll Nutr 1998;17:322–326.

58 Milne DB, Nielsen FH: Effects of a diet low in copper on copper-status indicators in post-menopausal women. Am J Clin Nutr 1996;63:358–364.

59 Medeiros DM, Nielsen FH: Effect of a diet low in copper on copper-status indicators in post-menopausal women. Am J Clin Nutr 1996;63:358–364.

60 Turley E, McKeown A, Bonham MP, O'Connor JM, Chopra M, Harvey LJ, Majsak-Newman G, Fairweather-Tait SJ, Bugel S, Sandstrom B, Pock E, Mazur A, Rayssiguier Y, Strain JJ: Copper supplementation in humans does not affect the susceptibility of low density lipoprotein to in vitro induced oxidation (FOODCUE project). Free Radic Biol Med 2000;29:1129–1134.

61 Rock E, Mazur A, O'Connor JM, Bonham MP, Rayssiguier Y, Strain JJ: The effect of copper supplementation on red blood cell oxidizability and plasma antioxidants in middle-aged healthy volunteers. Free Radic Biol Med 2000;28:324–329.

62 Conlan D, Korula R, Tallentire D: Serum copper levels in elderly patients with femoral-neck fractures. Age Ageing 1990;19:212–214.

63 Eaton-Evans J, Mellwrath EM, Jackson WE, McCartney H, Strain JJ: Copper supplementation and the maintenance of bone mineral density in middle-aged women. J Trace Elem Exp Med 1996;9:87–94.

64 Baker A, Harvey L, Majask-Newman G, Fairweather-Tait S, Flynn A, Cashman K: Effect of dietary copper intakes on biochemical markers of bone metabolism in healthy adult males. Eur J Clin Nutr 1999;53:408–412.

65 Baker A, Turley E, Bonham MP, O'Connor JM, Strain JJ, Flynn A, Cashman KD: No effect of copper supplementation on biochemical markers of bone metabolism in healthy adults. Br J Nutr 1999;82:283–290.

66 Percival SS: Copper and immunity. Am J Clin Nutr 1998;67(suppl):1064–1068.

67 Failla ML, Hopkins RG: Is low copper status immunosuppressive? Nutr Rev 1998;56:S59–S64.

68 Heresi G, Castillo-Duran C, Munoz C, Arevalo M, Schlesinger L: Phagocytosis and immunoglob-ulin levels in hypocupremic children. Nutr Res 1985;5:1327–1334.

69 Kelley DS, Daudu PA, Taylor PC, Makey BE, Turnlund JR: Effects of low-copper diets on human immune response. Am J Clin Nutr 1995;62:412–416.

70 Bremner I: Manifestations of copper excess. Am J Clin Nutr 1998;67(suppl):1069–1073.

71 Fitzgerald DJ: Safety guidelines for copper in water. Am J Clin Nutr 1998;67(suppl):1098–1102.

72   Burk RF, Levander OA: Selenium; in Shils M, Olson JA, Shike M, Ross AC (eds): Nutrition in Health and Disease, ed 9. Baltimore, Williams & Wilkins, 1999, pp 265–276.

73   Rayman MP: The importance of selenium to human health. Lancet 2000;356:233–241.

74   Chang JC: Selenium content of brazil nuts from two geographic locations in Brazil. Chemosphere 1995;30:801–802.

75   Holben DH, Smith AM: The diverse role of selenium within selenoproteins: A review. J Am Diet Assoc 1999;99:836–843.

76   Ursini F, Heim S, Kiess M, Maiorino M, Roven A, Wissing J, Flohe L: Dual function of the selenoprotein PHGPx during sperm maturation. Science 1999;285:1393–1396.

77   Mustacich D, Powis G: Thioredoxin reductase. Biochem J 2000;346:1–8.

78   Larsen PR, Davies TF, Hay ID: The thyroid gland; in Wilson JD, Foster DW, Kronenberg HM, Larsen PR (eds): Williams Textbook of Endocrinology, ed 9. Philadelphia, Saunders, 1998, pp 389–515.

79   Arteel GE, Briviba K, Sies H: Protein against peroxynitrite. FEBS Lett 1999;445:226–230.

80   Levander OA: Coxsackievirus as a model of viral evolution driven by dietary oxidative stress. Nutr Rev 2000;58:S17–S24.

81   Food and Nutrition Board, Institute of Medicine: Selenium. Dietary Reference Intakes for Vitamin C, Vitamin E, Selenium, and Carotenoids. Washington, National Academy Press, 2000, pp 284–324.

82   Foster LH, Sumar S: Selenium in health and disease: A review. Crit Rev Food Sci Nutr 1997; 37:211–228.

83   Salonen JT, Alfthan G, Huttunen JK, Pikkarainen J, Puska P: Association between cardiovascular death and myocardial infarction and serum selenium in a matched-pair longitudinal study. Lancet 1982;ii:175–179.

84   Virtamo J, Valkeila E, Alfthan G, Punsar S, Huttunen JK, Karvonen MJ: Serum selenium and risk of coronary heart disease in stroke. Am J Epidemiol 1985;122:276–282.

85   Suadicani P, Hein HO, Gyntelberg F: Serum selenium concentration and risk of ischaemic heart disease in a prospective cohort study of 3,000 males. Atherosclerosis 1992;96:33–42.

86   Salvini S, Hennekens CH, Morris JS, Willett WC, Stampfer MJ: Plasma levels of the antioxidant selenium and risk of myocardial infarction among US physicians. Am J Cardiol 1995;76: 1218–1221.

87   Kardinaal AF, Kok FJ, Kohlmeier L, Martin-Moreno JM, Ringstad J, Gomez-Aracena J, Mazaev VP, Thamm M, Martin BC, Aro A, Kark JD, Delgado-Rodriguez M, Riemersma RA, Van't Veer P: Association between risk of acute myocardial infarction in European men. The EURAMIC Study. European Antioxidant Myocardial Infarction and Breast Cancer. Am J Epidemiol 1997;145:373–379.

88   Rayman MP, Clark LC: Selenium in cancer prevention; in Roussel AM (ed): Trace Elements in Man and Animals, ed 10. New York, Plenum Press, 2000, pp 575–580.

89   Combs GF Jr, Gray WP: Chemopreventive agents: Selenium. Pharmacol Ther 1998;79: 179–192.

90   Garland M, Morris JS, Stampfer MJ, Colditz GA, Spate VL, Baskett CK, Rosner B, Speizer FE, Willett WC, Hunter DJ: Prospective study of toenail selenium levels and cancer among women. J Natl Cancer Inst 1995;87:497–505.

91   Yu MW, Horng IS, Hsu KH, Chiang YC, Liaw YF, Chen CJ: Plasma selenium levels and risk of hepatocellular carcinoma among men with chronic hepatitis virus infection. Am J Epidemiol 1999;150:367–374.

92   Knekt P, Marniemi J, Teppo L, Heliovaara M, Aromaa A: Is low selenium status a risk factor for lung cancer? Am J Epidemiol 1998;148:975–982.

93   Yoshizawa K, Willett WC, Morris SJ, Stampfer MJ, Spiegelman D, Rimm EB, Giovannucci E: Study of prediagnostic selenium level in toenails and the risk of advanced prostate cancer. J Natl Cancer Inst 1998;90:1219–1224.

94   Nomura AM, Lee J, Stemmermann GN, Combs GF Jr: Serum selenium and subsequent risk of prostate cancer. Cancer Epidemiol Biomarkers Prev 2000;9:883–887.

95   Brooks JD, Metter EJ, Chan DW, Sokoll LJ, Landis P, Nelson WG, Muller D, Andres R, Carter HB: Plasma selenium level before diagnosis and the risk of prostate cancer development. J Urol 2001; 166:2034–2038.

96   Ghadirian P, Maissonneuve P, Perret C, Kennedy G, Boyle P, Krewski D, Lacroix A: A case-control study of toenail selenium and cancer of the breast, colon, and prostate. Cancer Detect Prev 2000; 24:305–313.

97   Yu SY, Zhu YJ, Li WG: Protective role of selenium against hepatitis B virus and primary liver cancer in Qidong. Biol Trace Elem Res 1997;56:117–124.

98   Duffield-Lillico AJ, Dalkin BL, Reid ME, Turnbull W, Slate EH, Jacobs ET, Marshall JR, Clark LC: Selenium supplementation, baseline plasma selenium status and incidence of prostate cancer: An analysis of the complete treatment period of the Nutritional Prevention of Cancer Trial. BJU Int 2003;91:608–612.

99   Duffield-Lillico AJ, Slate EH, Reid ME, Turnbull W, Wilkins PA, Combs GF Jr, Park HK, Gross EG, Graham GF, Stratton MS, Marshall JR, Clark LC: Selenium supplementation and secondary prevention of nonmelanoma skin cancer in a randomized trial. J Natl Cancer Inst 2003;95:1477–1481.

100   Clark LC, Marshall JR: Randomized, controlled chemoprevention trials in populations at very high risk for prostate cancer: Elevated prostate-specific antigen and high-grade prostatic intraepithelial neoplasia. Urology 2001;57(suppl 1):185–187.

101   Klein EA, Thompson IM, Lippman SM, Goodman PJ, Albanes D, Taylor PR, Coltman C: SELECT – The Selenium and Vitamin E Cancer Prevention Trial: Rationale and design. Prostate Cancer Prostatic Dis 2000;3:145–151.

102   Roy M, Kiremidjian-Schumacher L, Wishe HI, Cohen MW, Stotzky G: Supplementation with selenium and human immune cell functions. II. Effect on lymphocyte proliferation and interleukin-2 receptor expression. Biol Trace Elem Res 1994;41:103–114.

103   Kiremidjian-Schumacher L, Roy M, Whishe HI, Cohen MW, Stotzky G: Supplementation with selenium and human immune cell functions. II. Effect on cytotoxic lymphocytes and natural killer cells. Biol Trace Elem Res 1994;41:115–127.

104   Kiremidjian-Schumacher L, Roy M, Glickman R, Schneider K, Rothstein S, Cooper J, Hochster H, Kim M, Newman R: Selenium and immunocompetence in patients with head and neck cancer. Biol Trace Elem Res 2000;73:97–111.

105   Baum MK, Miguez-Burbano MJ, Campa A, Shor-Posner G: Selenium and interleukins in persons infected with human immunodeficiency virus type I. J Infect Dis 2000;182 (suppl 1):69–73.

106   Beck MA, Esworthy RS, Ho YS, Chu FF: Glutathione peroxidase protects mice from viral-induced myocarditis. Faseb J 1998;12:1143–1149.

107   Prasad AS, Halsted JA, Nadimi M: Syndrome of iron deficiency anemia, hepatosplenomegaly, hypogonadism, dwarfism, and geophagia. Am J Med 1961;31:532–546.

108   Prasad AS: Zinc deficiency in humans: A neglected problem. J Am Coll Nutr 1998;17:542–543.

109   King JC, Keen CL: Zinc; in Shils M, Olson JA, Shike M, Ross AC (eds): Nutrition in Health and Disease, ed 9. Baltimore, Williams & Wilkins, 1999, pp 223–239.

110   Food and Nutrition Board, Institute of Medicine: Zinc. Dietary Reference Intakes for Vitamin A, Vitamin K, Boron, Chromium, Copper, Iodine, Iron, Manganese, Molybdenum, Nickel, Silicon, Vanadium, and Zinc. Washington, National Academy Press, 2001, pp 442–501.

111   Cousins RJ: Zinc; in Ziegler EE, Filer LJ (eds): Present Knowledge in Nutrition. Washington, ILSI Press, 1996, pp 293–306.

112   O'Dell BL: Role of zinc in plasma membrane function. J Nutr 2000;130(suppl):1432–1436.

113   Truong-Tran AQ, Ho LH, Chai F, Zalewski PD: Cellular zinc fluxes and the regulation of apoptosis/gene-directed cell death. J Nutr 2000;130(suppl):1459–1466.

114   O'Brian KO, Zavaleta N, Caulfield LE, Wen J, Abrams SA: Prenatal iron supplements impair zinc absorption in pregnant Peruvian women. J Nutr 2000;130:2251–2255.

115   Fung EB, Ritchie LD, Woodhouse LR, Roehl R, King JC: Zinc absorption in women during pregnancy and lactation: A longitudinal study. Am J Clin Nutr 1997;66:80–88.

116   Wood RJ, Zheng JJ: High dietary calcium intakes reduce zinc absorption and balance in humans. Am J Clin Nutr 1997;65:1803–1809.

117   McKenna AA, Ilich JZ, Andon MB, Wang C, Matkovic V: Zinc balance in adolescent females consuming a low- or high-calcium diet. Am J Clin Nutr 1997;65:1460–1464.

118   Kauwell GP, Bailey LB, Gregory JR 3rd, Bowling DW, Cousins RJ: Zinc status is not adversely affected by folic acid supplementation and zinc intake does not impair folate utilization in human subjects. J Nutr 1995;125:66–72.

119 Hambridge M: Human zinc deficiency. J Nutr 2000;130(suppl):1355–1349.

120 Walravens PA, Hambridge KM, Koepfer DM: Zinc supplementation in infants with a nutritional pattern of failure to thrive: A double-blind, controlled study. Pediatrics 1989;83:532–538.

121 Hambridge M, Krebs N: Zinc and growth; in Roussel AM (ed): Trace Elements in Man and Animals. Proc 10th Int Symp on Trace Elements in Man and Animals. New York, Plenum Press, 2000, pp 977–980.

122 MacDonald RS: The role of zinc in growth and cell proliferation. J Nutr 2000;130:1500–1508.

123 Caulfield LE, Zavaleta N, Shankar AH, Merialdi M: Potential contribution of maternal zinc supplementation during pregnancy to maternal and child survival. Am J Clin Nutr 1998;68:499–508.

124 Sandstead HH, Penland JG, Alcock NW, Dayal HH, Chen XC, Li JS, Zhao F, Yang JJ: Effects of repletion with zinc and other micronutrients on neuropsychologic performance and growth of Chinese children. Am J Clin Nutr 1998;68:470–475.

125 Black MM: Zinc deficiency and child development. Am J Clin Nutr 1998;68:464–469.

126 Baum MK, Shor-Posner G, Campa A: Zinc status in human immunodeficiency virus infection. J Nutr 2000;130:1421–1423.

127 Shankar AH, Prasad AS: Zinc and immune function: The biological basis of altered resistance to infection. Am J Clin Nutr 1998;68:447–463.

128 Salgueiro MJ, Zubillaga M, Lysionek A, Cremaschi G, Goldman CG, Caro R, De Paoli T, Hager A, Weill R, Boccio J: Zinc status and immune system relationship: A review. Biol Trace Elem Res 2000;76:193–205.

129 Fortes C, Forastiere F, Agabiti N, Fano V, Pacifici R, Virgili F, Piras G, Guidi L, Bartolini C, Tricerri A, Zuccano P, Ebrahim S, Perucci A: The effect of zinc and vitamin A supplementation on immune response in an older population. J Am Geriatr Soc 1998;46:19–26.

130 Caulfield LE, Zavaleta N, Figueroa A, Leon Z: Maternal zinc supplementation does not affect size at birth or pregnancy duration in Peru. J Nutr 1999;129:1563–1568.

131 Osendarp SJ, van Raaij JM, Arifeen SE, Wahed M, Baqui AH, Fuchs GJ: A randomized, placebo-controlled trial of the effect of zinc supplementation during pregnancy on pregnancy outcome in Bangladeshi urban poor. Am J Clin Nutr 2000;71:114–119.

132 VandenLangenberg GM, Mares-Perlman JA, Klein R, Klein BE, Brady WE, Palta M: Associations between antioxidant and zinc intake and the 5-year incidence of early age-related maculopathy in the Beaver Dam Eye Study. Am J Epidemiol 1998;148:204–214.

133 Smith W, Mitchell P, Webb K, Leeder SR: Dietary antioxidants and age-related maculopathy. The Blue Mountains Eye Study. Ophthalmology 1999;106:761–767.

134 Cho E, Stampfer MJ, Seddon JM, Hung S, Spiegelman D, Rimm EB, Willett WC, Hankinson SE: Prospective study of zinc intake and the risk of age-related macular degeneration. Ann Epidemiol 2001;11:328–336.

135 Newsome DA, Swartz M, Leone NC, Elston RC, Miller E: Oral zinc in macular degeneration. Arch Ophthalmol 1988;106:192–198.

136 Stur M, Tittl M, Reitner A, Meisinger V: Oral zinc and the second eye in age-related macular degeneration. Invest Ophthalmol Vis Sci 1996;37:1225–1235.

137 AREDS Report No 8: A randomized, placebo-controlled, clinical trial of high-dose supplementation with vitamins C and E, β-carotene, and zinc for age-related macular degeneration and vision loss. Arch Ophthalmol 2001;119:1417–1436.

138 Evans JR: Antioxidant vitamin and mineral supplements for age-related macular degeneration (Cochrane Review). Cochrane Database Syst Rev 2002:CD000254.

139 International Agency for Research on Cancer; IARC Handbooks of Cancer Prevention: Carotenoids. Lyon, International Agency for Research on Cancer, 1998.

140 Food and Nutrition Board, Institute of Medicine. Appendix C: Dietary Intake Data from the Third National Health and Nutrition Examination Survey (NHANES III), 1988–1994. Dietary Reference Intakes for Vitamin A, Vitamin K, Arsenic, Boron, Chromium, Copper, Iodine, Iron, Manganese, Molybdenum, Nickel, Silicon, Vanadium, and Zinc. Washington, National Academy Press, 2001, pp 594–605.

141 Clinton SK: Lycopene: Chemistry, biology, and implications for human health and disease. Nutr Rev 1998;56:35–51.

142 Food and Nutrition Board, Institute of Medicine: β-Carotene and Other Carotenoids. Dietary Reference Intakes for Vitamin C, Vitamin E, Selenium, and Carotenoids. Washington, National Academy Press, 2000, pp 325–400.

143 Halliwell B, Gutteridge JMC: Free Radicals in Biology and Medicine, ed 3. New York, Oxford University Press, 1999.

144 Di Mascio P, Kaiser S, Sies H: Lycopene as the most efficient biological carotenoid singlet oxygen quencher. Arch Biochem Biophys 1989;274:532–538.

145 Young AJ, Lowe GM: Antioxidant and prooxidant properties of carotenoids. Arch Biochem Biophys 2001;385:20–27.

146 Krinsky NI, Landrum JT, Bone RA: Biological mechanisms of the protective role of lutein and zeaxanthin in the eye. Annu Rev Nutr 2003;23:171–201.

147 Bertram JS: Carotenoids and gene regulation. Nutr Rev 1999;57:182–191.

148 Stahl W, Nicolai S, Briviba K, Hanusch M, Broszeit G, Peters M, Martin HD, Sies H: Biological activities of natural and synthetic carotenoids: Induction of gap junctional communication and singlet oxygen quenching. Carcinogenesis 1997;18:89–92.

149 Van Poppel G, Spanhaak S, Ockhuizen T: Effect of β-carotene on immunological indexes in healthy male smokers. Am J Clin Nutr 1993;57:402–407.

150 Hughes DA, Wright AJ, Finglas PM, Peerless AC, Bailey AL, Astley SB, Pinder AC, Southon S: The effect of β-carotene supplementation on the immune function of blood monocytes from healthy male non-smokers. J Lab Clin Med 1997;129:309–317.

151 Santos MS, Gaziano JM, Leka LS, Beharka AA, Hennekens CH, Meydani SN: β-Carotene-induced enhancement of natural killer cell activity in elderly men: An investigation of the role of cytokines. Am J Clin Nutr 1998;68:164–170.

152 Watzl B, Bub A, Blockhaus M, Herbert BM, Luhrmann PM, Neuhauser-Berthold M, Rechkemmer G: Prolonged tomato juice consumption has no effect on cell-mediated immunity of well-nourished elderly men and women. J Nutr 2000;130:1719–1723.

153 Kritchevsky SB: β-Carotene, carotenoids and the prevention of coronary heart disease. J Nutr 1999;129:5–8.

154 Rissanen TH, Voutilainen S, Nyyssonen K, Salonen R, Kaplan GA, Salonen JT: Serum lycopene concentrations and carotid atherosclerosis: The Kuopio Ischaemic Heart Disease Risk Factor Study. Am J Clin Nutr 2003;77:133–138.

155 Dwyer JH, Paul-Labrador MJ, Fan J, Shircore AM, Merz CN, Dwyer KM: Progression of carotid intima-media thickness and plasma antioxidants: The Los Angeles Atherosclerosis Study. Arterioscler Thromb Vasc Biol 2004;24:313–319.

156 McQuillan BM, Hung J, Beilby JP, Nidorf M, Thompson PL: Antioxidant vitamins and the risk of carotid atherosclerosis. The Perth Carotid Ultrasound Disease Assessment Study (CUDAS). J Am Coll Cardiol 2001;38:1788–1794.

157 Rissanen T, Voutilainen S, Nyyssonen K, Salonen R, Salonen JT: Low plasma lycopene concentration is associated with increased intima-media thickness of the carotid artery wall. Arterioscler Thromb Vasc Biol 2000;20:2677–2681.

158 D'Odorico A, Martines D, Kiechl S, et al: High plasma levels of α- and β-carotene are associated with a lower risk of atherosclerosis: Results from the Bruneck study. Atherosclerosis 2000;153: 231–239.

159 Iribarren C, Folsom AR, Jacobs DR Jr, Gross MD, Belcher JD, Eckfeldt JH: Association of serum vitamin levels, LDL susceptibility to oxidation, and autoantibodies against MDA-LDL with carotid atherosclerosis. A case-control study. The ARIC Study Investigators. Atherosclerosis Risk in Communities. Arterioscler Thromb Vasc Biol 1997;17:1171–1177.

160 Sesso HD, Buring JE, Norkus EP, Gaziano JM: Plasma lycopene, other carotenoids, and retinol and the risk of cardiovascular disease in women. Am J Clin Nutr 2004;79:47–53.

161 Rissanen TH, Voutilainen S, Nyyssonen K, Lakka TA, Sivenius J, Salonen R, Kaplan GA, Salonen JT: Low serum lycopene concentration is associated with an excess incidence of acute coronary events and stroke: The Kuopio Ischaemic Heart Disease Risk Factor Study. Br J Nutr 2001;85:749–754.

162 Street DA, Comstock GW, Salkeld RM, Schuep W, Klag MJ: Serum antioxidants and myocardial infarction. Are low levels of carotenoids and α-tocopherol risk factors for myocardial infarction? Circulation 1994;90:1154–1161.

163 Hak AE, Stampfer MJ, Campos H, Sesso HD, Gaziano JM, Willett WC, Ma J: Plasma carotenoids and tocopherols and risk of myocardial infarction in a low-risk population of US male physicians. Circulation 2003;108:802–807.

164 Evans RW, Shaten BJ, Day BW, Kuller LH: Prospective association between lipid soluble antioxidants and coronary heart disease in men. The Multiple Risk Factor Intervention Trial. Am J Epidemiol 1998;147:180–186.

165 Sahyoun NR, Jacques PF, Russell RM: Carotenoids, vitamins C and E, and mortality in an elderly population. Am J Epidemiol 1996;144:501–511.

166 Rimm EB, Stampfer MJ, Ascherio A, Giovannucci E, Colditz GA, Willett WC: Vitamin E consumption and the risk of coronary heart disease in men. N Engl J Med 1993;328: 1450–1456.

167 Gaziano JM, Manson JE, Brnch LG, Colditz GA, Willett WC, Buring JE: A prospective study of consumption of carotenoids in fruits and vegetables and decreased cardiovascular mortality in the elderly. Ann Epidemiol 1995;5:255–260.

168 Osganian SK, Stampfer MJ, Rimm E, Spiegelman D, Manson JE, Willett WC: Dietary carotenoids and risk of coronary artery disease in women. Am J Clin Nutr 2003;77:1390–1399.

169 α-Tocopherol, β-Carotene Cancer Prevention Study Group: The effect of vitamin E and β-carotene on the incidence of lung cancer and other cancers in male smokers. N Engl J Med 1994;330:1029–1035.

170 Hennekens CH, Buring JE, Manson JE, Stampfer M, Rosner B, Cook NR, Belanger C, LaMotte F, Gaziano JM, Ridker PM, Willett WC, Peto R: Lack of effect of long-term supplementation with β-carotene on the incidence of malignant neoplasms and cardiovascular disease. N Engl J Med 1996;334:1145–1149.

171 Greenberg ER, Baron JA, Karagas MR, Stukel TA, Nierenberg DW, Stevens MM, Mandel JS, Halle W: Mortality associated with low plasma concentration of β-carotene and the effect of oral supplementation. JAMA 1996;275:699–703.

172 Omenn GS, Goodman GE, Thornquist MD, Balmes J, Cullen MR, Glass A, Keogh JP, Meyskens FL, Valanis B, Williams JH, Barnhart S, Hammar S: Effects of a combination of β-carotene and vitamin A on lung cancer and cardiovascular disease. N Engl J Med 1996;334: 1150–1155.

173 US Preventive Services Task Force: Routine vitamin supplementation to prevent cancer and cardiovascular disease: Recommendations and rationale. Ann Intern Med 2003;139:51–55.

174 Morris CD, Carson S: Routine vitamin supplementation to prevent cardiovascular disease: A summary of the evidence for the US Preventive Services Task Force. Ann Intern Med 2003;139:56–70.

175 Mares-Perlman JA, Millen AE, Ficek TL, Hankinson SE: The body of evidence to support a protective role for lutein and zeaxanthin in delaying chronic disease. Overview. J Nutr 2002;132: 518–524.

176 Snellen EL, Verbeek AL, Van Den Hoogen GW, Cruysberg JR, Hoying CB: Neovascular age-related macular degeneration and its relationship to antioxidant intake. Acta Ophthalmol Scand 2002;80:368–371.

177 Mares-Perlman JA, Fisher AI, Klein R, Palta M, Block G, Millen AE, Wrigt JD: Lutein and zeaxanthin in the diet and serum and their relation to age-related maculopathy in the Third National Health and Nutrition Examination Survey. Am J Epidemiol 2001;153:424–432.

178 Seddon JM, Ajani UA, Sperduto RD, Hiller R, Blair N, Burton TC, Farber MD, Gragoudas ES, Hallar J, Miller DT, et al: Dietary carotenoids, vitamins A, C, and E, and advance age-related macular degeneration. Eye Disease Case-Control Study Group. JAMA 1994;272:1413–1420.

179 Gale CR, Hall NF, Phillips DI, Martyn CN: Lutein and zeaxanthin status and risk of age-related macular degeneration. Invest Ophthalmol Vis Sci 2003;44:2461–2465.

180 Eye Disease Case-Control Study Group: Antioxidant status and neovascular age-related macular degeneration. Arch Ophthalmol 1993;111:104–109.

181 Bone RA, Landrum JT, Mayne ST, Gomez CM, Tibor SE, Twaroska EE: Macular pigment in donor eyes with and without AMD: A case-control study. Invest Ophthalmol Vis Sci 2001;42:235–240.

182 Beatty S, Murray IJ, Henson DB, Carden D, Koh H, Bouton ME: Macular pigment and risk for age-related macular degeneration in subjects from a northern European population. Invest Ophthalmol Vis Sci 2001;42:439–446.

183   Flood V, Smith W, Wang JJ, Manzil F, Webb K, Michell P: Dietary and antioxidant intake and incidence of early age-related maculopathy: The Blue Mountains Eye Study. Ophthalmology 2002; 109:2272–2278.

184   Mares-Perlman JA, Klein R, Klein BE, Greger JL, Brady WE, Palta M, Ritter LL: Association of zinc and antioxidant nutrients with age-related maculopathy. Arch Ophthalmol 1996;114:991–997.

185   Mares-Perlman JA, Brady WE, Klein R, Klein BE, Bowen P, Stacewicz-Sapuntzakis M, Palta M: Serum antioxidants and age-related macular degeneration in a population-based case-control study. Arch Ophthalmol 1995;113:1518–1523.

186   Mares-Perlman JA: Too soon for lutein supplements. Am J Clin Nutr 1999;70:431–432.

187   AREDS Report No 8: A randomized, placebo-controlled, clinical trial of high-dose supplementation with vitamins C and E, β-carotene, and zinc for age-related macular degeneration and vision loss. Arch Ophthalmol 2001;119:1417–1436.

188   Teikari JM, Laatikainen L, Virtamo J, Haukka J, Rautalahti M, Liesto K, Albanes D, Taylor P, Heinonen OP: Six-year supplementation with α-tocopherol and β-carotene and age-related maculopathy. Acta Ophthalmol Scand 1998;76:224–229.

189   Brown L, Rimm EB, Seddon JM, Giovannucci EL, Chasan-Taber L, Spiegelman D, Willett WC, Hankinson SE: A prospective study of carotenoid intake and risk of cataract extraction in US men. Am J Clin Nutr 1999;70:517–524.

190   Chasan-Taber L, Willett WC, Seddon JM, Stampfer MJ, Rosner B, Colditz GA, Speizer FE, Hankinson SE: A prospective study of carotenoid and vitamin A intakes and risk of cataract extraction in US women. Am J Clin Nutr 1999;70:509–516.

191   Lyle BJ, Mares-Perlman JA, Klein BE, Klein R, Greger JL: Antioxidant intake and risk of incident age-related nuclear cataracts in the Beaver Dam Eye Study. Am J Epidemiol 1999;149:801–809.

192   Christen WG, Manson JE, Glynn RJ, Gaziano JM, Sperduto RD, Buring JE, Hennekens CH: A randomized trial of β-carotene and age-related cataract in US physicians. Arch Ophthalmol 2003;121:372–378.

193   AREDS Report No 9: A randomized, placebo-controlled, clinical trial of high-dose supplementation with vitamins C and E and β-carotene for age-related cataract and vision loss. Arch Ophthalmol 2001;119:1439–1452.

194   Chylack LT Jr, Brown NP, Bron A, Hurst M, Kopcke W, Thiem U, Schalch W: The Roche European American Cataract Trial (REACT): A randomized clinical trial to investigate the efficacy of an oral antioxidant micronutrient mixture to slow progression of age-related cataract. Ophthalmic Epidemiol 2002;9:49–80.

195   Peto R, Doll R, Buckley JD, Sporn MB: Can dietary β-carotene materially reduce human cancer rates? Nature 1981;290:201–2108.

196   Ziegler RG: A review of epidemiologic evidence that carotenoids reduce the risk of cancer. J Nutr 1989;119:116–122.

197   Michaud DS, Feskanich D, Rimm EB, Colditz GA, Speizer FE, Willett WC, Giovannucci E: Intake of specific carotenoids and risk of lung cancer in two prospective US cohorts. Am J Clin Nutr 2000;72:990–997.

198   Holick CN, Michaud DS, Stolzenberg-Solomon R, Mayne ST, Pietinen P, Taylor PR, Virtamo J, Albanes D: Dietary carotenoids serum β-carotene, and retinol and risk of lung cancer in the α-tocopherol, β-carotene cohort study. Am J Epidemiol 2002;156:536–547.

199   Voorrips LE, Goldbohm RA, Brants HA, Van Poppel GA, Sturmans F, Hermus RJ, Van den Brandt PA: A prospective cohort study on antioxidant and folate intake and male lung cancer risk. Cancer Epidemiol Biomarkers Prev 2000;9:357–365.

200   Mannisto S, Smith-Warner SA, Spiegelman D, Albanes D, Anderson K, Van den Brandt PA, Cerhan JR, Colditz G, Feskarich D, Freudenheim JL, Giovannucci E, Goldbohm RA, Graham S, Miller AB, Rohan TE, Virtamo J, Willett WC, Anater DJ: Dietary carotenoids and risk of lung cancer in a pooled analysis of seven cohort studies. Cancer Epidemiol Biomarkers Prev 2004; 13:40–48.

201   Omenn GS, Goodman GE, Thornquist MD, Balmes J, Cullen MR, Glass A, Keogh JP, Meyskens FL, Valanis B, Williams JH Jr, Barnhart S, Cherniack MG, Brodkin CH, Hammar S: Risk factors for lung cancer and for intervention effects in CARET, the β-Carotene and Retinol Efficacy Trial. J Natl Cancer Inst 1996;88:1550–1559.

202 Vainio H, Rautalahti M: An international evaluation of the cancer preventive potential of carotenoids. Cancer Epidemiol Biomarkers Prev 1998;7:725–728.

203 Giovannucci E: A review of epidemiologic studies of tomatoes, lycopene, and prostate cancer. Exp Biol Med (Maywood) 2002;227:852–859.

204 Giovannucci E, Ascherio A, Rimm EB, Stampfer MJ, Colditz GA, Willett WC: Intake of carotenoids and retinol in relation to risk of prostate cancer. J Natl Cancer Inst 1995;87:1767–1776.

205 Mills PK, Beeson WL, Phillips RL, Fraser GE: Cohort study of diet, lifestyle, and prostate cancer in Adventist men. Cancer 1989;64:598–604.

206 Gann PH, Ma J, Giovannucci E, Willett WC, Sacks FM, Henekens CH, Stampfer MJ: Lower prostate cancer risk in men with elevated plasma lycopene levels: Results of a prospective analysis. Cancer Res 1999;59:1225–1230.

207 Schuurman AG, Goldbohm RA, Brants HA, van den Brandt PA: A prospective cohort study on intake of retinol, vitamins C and E, and carotenoids and prostate cancer risk (Netherlands). Cancer Causes Control 2002;13:573–582.

208 Van Het Hof KH, West CE, Weststrate JA, Hautvast JG: Dietary factors that affect the bioavailability of carotenoids. J Nutr 2000;130:503–506.

209 Van Het Hof KH, Brouwer IA, West CE, Haddeman E, Steegers-Theunissen RP, Van Dusseldorp M, Wertstrate JA, Eskes TK, Hantrast JG: Bioavailability of lutein from vegetables is five times higher than that of β-carotene. Am J Clin Nutr 1999;70:261–268.

210 Gartner C, Stahl W, Sies H: Lycopene is more bioavailable from tomato paste than from fresh tomatoes. Am J Clin Nutr 1997;66:116–122.

211 Stahl W, Sies H: Uptake of lycopene and its geometrical isomers is greater from heat-processed than from unprocessed tomato juice in humans. J Nutr 1992;122:2161–2166.

212 β-Carotene: Natural medicines comprehensive database. http://www.natural data base.com

213 Ehrenfeld M, Levy M, Sharon P, Rachmilewitz D, Eliakim M: Gastrointestinal effects of long-term colchicines therapy in patients with recurrent polyserositis (familial Mediterranean fever). Dig Dis Sci 1982;27:723–727.

214 Tang G, Serfaty-Lacrosniere C, Camilo ME, Russell RM: Gastric acidity influences the blood response to a β-carotene dose in humans. Am J Clin Nutr 1996;64:622–626.

215 Van den Berg H: Carotenoid interactions. Nutr Rev 1999;57:1–10.

216 Kostic D, White WS, Olson JA: Intestinal absorption, serum clearance, and interactions between lutein and β-carotene when administered to human adults in separate or combined oral doses. Am J Clin Nutr 1995;62:604–610.

217 Albanes D, Cirtamo J, Taylor PR, Rautalahti M, Pietinen P, Heinonen OP: Effects of supplemental β-carotene, cigarette smoking, and alcohol consumption on serum carotenoids in the α-Tocopherol, β-Carotene Cancer Prevention Study. Am J Clin Nutr 1997;66:366–372.

218 Nierenberg DW, Dain BJ, Mott LA, Baron JA, Greenberg ER: Effects of 4 years of oral supplementation with β-carotene on serum concentrations of retinol, tocopherol, and five carotenoids. Am J Clin Nutr 1997;66:315–1319.

219 Wahlqvist ML, Wattanapenpaiboon N, Macrae FA, Lambert JR, MacLennan R, Hsu-Hage BH: Changes in serum carotenoids in subjects with colorectal adenomas after 24 months of β-carotene supplementation. Australian Polyp Prevention Project Investigators. Am J Clin Nutr 1994;60: 936–943.

220 Mayne ST, Cartmel B, Silva F, Kim CS, Fallon BG, Briskin K, Zheng T, Baum M, Shor-Posner G, Goodwin WJ Jr: Effect of supplemental β-carotene on plasma concentrations of carotenoids, retinol, and α-tocopherol in humans. Am J Clin Nutr 1998;68:642–647.

221 Harborne JB, Williams CA: Advances in flavonoid research since 1992. Phytochemistry 2000;55: 481–504.

222 Harborne JB: The comparative biochemistry of phytoalexin induction in plants. Biochem Syst Ecol 1999;27:335–368.

223 Li BQ, Fu T, Yan YD, Baylor NW, Ruscetti FW, Kung HF: Inhibition of HIV by baicalin. Cell Mol Biol Res 1997;39:119–124.

224 Harborne JB: Recent advances in chemical ecology. Nat Prod Rep 1999;16:509–523.

225 Arora A, Byrem TM, Nair MG, Strasburg GM: Modulation of liposomal membrane fluidity by flavonoids and isoflavonoids. Arch Biochem Biophys 2000;373:102–109.

226 Hodnick WF, Duval DL, Pardini RS: Inhibition of mitochondrial respiration and cyanide-stimulated generation of reactive oxygen species by selective flavonoids. Biochem Pharmacol 1994;47: 573–580.

227 Cos P, Ying L, Calomme M, Hu JP, Cimanga K, Van Poel B, Pieters L, Vlietinck AJ, Vanden BD: Structure-activity relationship and classification of flavonoids as inhibitors of xanthine oxidase and superoxide scavengers. J Nat Prod 1998;61:71–76.

228 Lale A, Herberg JM, Augereau JM, Billon M, LeConte M, Gleye J: Ability of different flavonoids to inhibit procoagulant activity of adherent human monocytes. J Nat Prod 1996;59: 273–276.

229 Melzig MF: Inhibition of adenosine deaminase activity of aortic endothelial cells by selected flavonoids. Planta Med 1996;62:20–21.

230 Rankin SM, de Whalley CV, Hoult JRS, Jessup W, Wikins GM, Collard J, Leake DS: The modification of low density lipoproteins by the flavonoids myricetin and gossypetin. Biochem Pharmacol 1993;45:67–75.

231 Frankel EN, Kanner J, German JB, Parks E, Kinsella JE: Inhibition of oxidation of human low-density lipoprotein by phenolic substances in red wine. Lancet 1993;341:454–456.

232 Leake DS: The possible role of antioxidants in fruits and vegetables in protecting against coronary heart disease; in Tomás-Barberau FA, Robin RJ (eds): Phytochemistry of Fruit and Vegetables. Oxford, Clarendon Press, 1997, pp 287–311.

233 Kavvadias D, Sand P, Youdim KA, Qaiser MZ, Rice-Evans C, Baur R, Sigel E, Rausch WD, Riederer P, Schreier P: The flavone hispidulin, a benzodiazepine receptor ligand with positive allosteric properties, traverses the blood-brain barrier and exhibits anti-convulsive effects. Br J Pharmacol 2004;142:811–820.

234 Hasrat JA, Pieters L, Claeys M, Vlietinck A: Adenosine-1 active ligands: Cirsimaritin, a flavone glycoside from *Microtea debilis*. J Nat Prod 1997;60:638–641.

235 Sekine T, Inagaki M, Ikegami F, Fuji Y, Ruangrungss N: Six diprenylisoflavones. Derrisisoflavones A–F, from *Derris scandens*. Phytochemistry 1998;52:87–94.

236 Liu L, Gitc DC, McClure JW: Effects of UV-B on flavonoids, ferulic acid, growth and photosynthesis in barley primary leaves. Physiol Plant 1995;93:725–733.

237 DFG Senate Commission on Food Safety (ed): Functional Food – Safety Aspects. Weinheim, Wiley-VCH, 2004.

238 Kramer K, Packer L: *R*-α-lipoic acid; in Kramer K, Hoppe P, Packer L (eds): Nutraceuticals in Health and Disease Prevention. New York, Dekker, 2001, pp 129–164.

239 Biewenga GP, Haenen GR, Bast A: The pharmacology of the antioxidant lipoic acid. Gen Pharmacol 1997;29:315–331.

240 Packer L, Kraemer K, Rimbach G: Molecular aspects of lipoic acid in the prevention of diabetes complications. Nutrition 2001;17:888–895.

241 Biewenga GP, Veening-Griffioen DH, Nicastia AJ, Haenen GR, Bast A: Effects of dihydrolipoic acid on peptide methionine sulfoxide reductase. Implications for antioxidant drugs. Arzneimittelforschung 1998;48:144–148.

242 Packer L: α-Lipoic acid: A metabolic antioxidant which regulates NF-κB signal transduction and protects against oxidative injury. Drug Metab Rev 1998;30:245–275.

243 Zhang WJ, Frei B: α-Lipoic acid inhibits TNF-α-induced NF-κB activation and adhesion molecule expression in human aortic endothelial cells. Faseb J 2001;15:2423–2432.

244 Mizuno M, Packer L: Effects of α-lipoic acid and dihydrolipoic acid on expression of proto-oncogene c-fos. Biochem Biophys Res Commun 1994;200:1136–1142.

245 Yeaman SJ, Kirby JA, Jones DE: Autoreactive responses to pyruvate dehydrogenase complex in the pathogenesis of primary biliary cirrhosis. Immunol Rev 2000;174:238–249.

246 Hagen TM, Ingersoll RT, Lykkesfeldt J, Liu J, Wehr CM, Vinarsky V, Bartholomew JC, Ames AB: (R)-α-lipoic acid-supplemented old rates have improved mitochondrial function, decreased oxidative damage, and increased metabolic rate. Faseb J 1999;13:411–418.

247 Suh JH, Shigeno ET, Morrow JD, Cox B, Rocha AE, Frei B, Hagen TM: Oxidative stress in the aging rat heart is reversed by dietary supplementation with (R)-(α)-lipoic acid. Faseb J 2001;15:700–706.

248 Hagen TM, Liu J, Lykkesfeldt J, Wehr CM, Ingersoll RT, Vinarsky V, Bartholomew JC, Ames BN: Feeding acetyl-*L*-carnitine and lipoic acid to old rats significantly improves metabolic function while decreasing oxidative stress. Proc Natl Acad Sci USA 2002;99:1870–1875.

249 Liu J, Head E, Gharib AM, Yuan W, Ingersoll RT, Hagen TM, Cotmann CW, Ames BN: Memory loss in old rats is associated with brain mitochondrial decay and RNA/DNA oxidation: Partial reversal by feeding acetyl-*L*-carnitine and/or *R*-α-lipoic acid. Proc Natl Acad Sci USA 2002; 99:2356–2361.

250 Bast A, Haenen GR: Lipoic acid: A multifunctional nutraceutical; in Kramer K, Hoppe P, Packer L (eds): Nutraceuticals in Health and Disease Prevention. New York, Dekker, 2001, pp 113–128.

251 Streeper RS, Henriksen EJ, Jacob S, Hokama JY, Fogt DL, Tritschler HJ: Differential effects of lipoic acid stereoisomers on glucose metabolism in insulin-resistant skeletal muscle. Am J Physiol 1997;273:E185–E191.

252 Estrada DE, Ewart HS, Tsakiridis T, Volchrunk A, Ramlal T, Tritschler H, Klip A: Stimulation of glucose uptake by the natural coenzyme α-lipoic acid/thioctic acid: Participation of elements of the insulin signalling pathway. Diabetes 1996;45:1798–1804.

253 Borcea V, Nourooz-Zadeh J, Wolff SP, Klevesath M, Hofmann M, Urich H, Wahl P, Ziegler R, Tritschler H, Halliwell B, Nawroth PP: α-Lipoic acid decreases oxidative stress even in diabetic patients with poor glycemic control and albuminuria. Free Radic Biol Med 1999;26:1495–1500.

254 Androne L, Gavan NA, Veresiu IA, Orasan R: In vivo effect of lipoic acid on lipid peroxidation in patients with diabetic neuropathy. In Vivo 2000;14:327–330.

255 Hofmann MA, Schiekofer S, Kanitz M, Klevesath MS, Joswig M, Lee V, Morcos M, Tritschler H, Ziegler R, Wahl P, Bierhaus A, Nawroth PP: Insufficient glycemic control increases nuclear factor-κB binding activity in peripheral blood mononuclear cells isolated from patients with type 1 diabetes. Diabetes Care 1998;21:1310–1316.

256 Biewenga G, Haenen GR, Bast A: The role of lipoic acid in the treatment of diabetic polyneuropathy. Drug Metab Rev 1997;29:1025–1054.

257 Diabetes Control and Complications Trial Research Group: The effect of intensive treatment of diabetes on the development and progression of long-term complications in insulin-dependent diabetes mellitus. N Engl J Med 1993;329:977–986.

258 Ziegler D, Hanefeld M, Ruhnau KJ, Meissner HP, Lobisch M, Schutte K, Gries FA: Treatment of symptomatic diabetic peripheral neuropathy with the antioxidant α-lipoic acid. A 3-week multicentre randomized controlled trial (ALADIN Study). Diabetologia 1995;38:1425–1433.

259 Ziegler D, Hanefeld M, Ruhnau KJ, Hasche H, Lobisch M, Schutte K, Kerum G, Malessas R: Treatment of symptomatic diabetic polyneuropathy with the antioxidant α-lipoic acid: A 7-month multicentre randomized controlled trial (ALADIN III Study). ALADIN III Study Group. α-Lipoic Acid in Diabetic Neuropathy. Diabetes Care 1999;22:1296–1301.

260 Reljanovic M, Reichel G, Rett K, Lobisch M, Schutte K, Moller W, Tritschler HJ, Mehnert H: Treatment of diabetic polyneuropathy with the antioxidant thioctic acid (α-lipoic acid): A two-year multicentre randomized double-blind placebo-controlled trial (ALADIN II). A Lipoic Acid in Diabetic Neuropathy. Free Radic Res 1999;31:171–179.

261 Heitzer T, Finckh B, Albers S, Krohn K, Kohlschutter A, Meinertz T: Beneficial effects of α-lipoic acid and ascorbic acid on endothelium-dependent, nitric oxide-mediated vasodilation in diabetic patients: Relation to parameters of oxidative stress. Free Radic Biol Med 2001;31: 53–61.

262 Haak E, Usadel KH, Kusterer K, Amini P, Frommeyer R, Tritschler HJ, Haak T: Effects of α-lipoic acid on microcirculation in patients with peripheral diabetic neuropathy. Exp Clin Endocrinol Diabetes 2000;108:168–174.

263 Morcos M, Borcea V, Isermann B, Gehrke S, Ehret H, Henkels M, Schiekofer S, Hofmann M, Amiral J, Tritschler H, Ziegler R, Wahl P, Nawroth PP: Effect of α-lipoic acid on the progression of endothelial cell damage and albuminuria in patients with diabetes mellitus: An exploratory study. Diabetes Res Clin Pract 2001;52:175–183.

264 Hermann R, Niebch G, Borbe HO: Enantioselective pharmacokinetics and bioavailability of different racemic α-lipoic acid formulations in healthy volunteers. Eur J Pharm Sci 1996; 167–174.

265   Maitra I, Serbinova E, Tritschler HJ, Packer L: Stereospecific effects of *R*-lipoic acid on buthionine sulfoximine-induced cataract formation in new-born rats. Biochem Biophys Res Commun 1996;221:422–429.
266   Hendler SS, Rovik DR (eds): PDR for Nutritional Supplements. Montvale, Medical Economics Co Inc, 2001.

Peter Schreier
Lehrstuhl für Lebensmittelchemie, Universität Würzburg
Am Hubland, DE–97074 Würzburg (Germany)
E-Mail schreier@pzlc.uni-wuerzburg.de

Augustin A (ed): Nutrition and the Eye.
Dev Ophthalmol. Basel, Karger, 2005, vol 38, pp 59–69

....

Vitamin C, Vitamin E and Flavonoids

K.M. Janisch, J. Milde, H. Schempp, E.F. Elstner

TUM Weihenstephan, Center of Life and Food Sciences, Lehrstuhl für Phytopathologie, Freising, Germany

Abstract

All inflammatory processes include oxygen-activating processes where reactive oxygen species are produced. Intrinsic radical scavenging systems or compounds administered with food warrant metabolic control within certain limits. Antioxidants, which in many cases are free radical scavengers or quenchers of activated states, comprise a vast number of classes of organic molecules including most prominently the phenolics. In this report, mechanisms of protection from oxidative damage by the antioxidants vitamin C and E and flavonoids, as present in most plant extracts used as natural drugs, are summarized. For this purpose the principle of oxygen activation during representative disease processes and the protective actions of antioxidants are outlined in short.

Reactive Oxygen Species and the Eye

Reactive oxygen species (ROS) occur in the healthy metabolism as by-products of oxidative processes such as the respiratory chain in mitochondria. Their detoxification is maintained with antioxidant enzymes (e.g. superoxide dismutase, catalase, glutathione peroxidase) and antioxidants (e.g. ascorbic acid, glutathione, tocopherol) and marks only small problems for a healthy individual. Certain incidences, such as ageing, inflammation, ischemia, and smoking, cause a rise in the oxidative stress. Thus the metabolism has to cope with ROS and ROS metabolism plays a crucial role in pathogenesis [1–3].

ROS include superoxide radicals ($O_2^{-\cdot}$), hydrogen peroxide (H_2O_2), hydroxyl radicals ($OH^{\cdot}$), singlet oxygen ($^1O_2$), peroxynitrite (ONNOH) and hypochloric acid (HOCl). A one-electron transfer onto oxygen generates superoxide radicals. They dismutate spontaneously to hydrogen peroxide and oxygen at neutral to slight acidic pH; whereas superoxide dismutases (SOD) catalyze

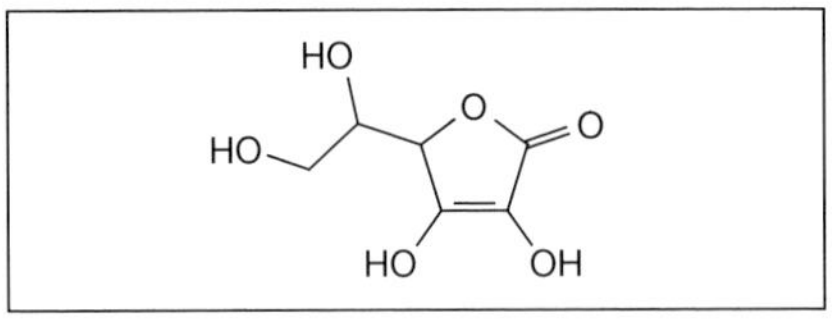

Fig. 1. Structure of ascorbic acid.

the dismutation independently from the pH. Hydrogen peroxide can also be generated via a two-electron transfer onto molecular oxygen, a reaction carried out by certain oxidases. Catalases detoxify hydrogen peroxide resulting in water and oxygen. A further one-electron transfer onto hydrogen peroxide produces the most reactive ROS, hydroxyl radicals. They react quickly with organic molecules in their surroundings generating further radicals such as alkyl, peroxyl radicals damaging the tissue with these interactions [1–4]. Singlet oxygen originates physically from the transfer of light energy onto molecular (triplet) oxygen through photosensitizers. It impairs organic molecules in the ground state [1, 4, 5].

ROS are known to be involved in many diseases and the process of ageing. In the eye, diseases such as age-related macular degeneration (AMD) and cataract are related to this. In both diseases, the high consumption of oxygen during normal metabolism and the exposure to light evolve oxidative stress in the tissue of the retina and lens [5]. In the course of AMD, phagocytic processes of the retinal pigment epithelial cells (RPE) and accumulation of lipofuscin, an age-related protein debris of the metabolism with photosensitizer abilities, enhance the oxidative strain in the retina additionally [5–8]. The degeneration of the lens fibers occurring in cataract is furthermore linked with smoking, cardiovascular diseases and diabetes [8], events already associated with the involvement of ROS. The protective mechanisms of the antioxidants vitamin C and E and flavonoids as well as their involvement in eye diseases are summarized in the following. Carotenoids are not considered here as they are discussed in detail in another chapter of this book.

Ascorbic Acid (Vitamin C)

The natural occurring vitamin ascorbic acid (vitamin C) is the most important hydrophilic antioxidant for human metabolism to maintain health. It is a dibasic acid with an enediol structure within the heterocyclic furanolactone ring (fig. 1) [9]. Ascorbic acid is able to chelate transition metals as a bidentate ligand but also possesses reducing abilities. It scavenges ROS (e.g. superoxide radical, hydroxyl radical, hydrogen peroxide, singlet oxygen) and is oxidized by a variety of other oxidants such as halogens, quinones, and phenoxyl radicals [9–12]. The oxidation of ascorbate to dehydroascorbate is a two-electron redox process. The loss of the two hydrogen ions and one electron forms the ascorbyl radical which is an acid and relatively stable compared to other radical ROS [9, 10]. The ascorbyl radicals disproportionate to ascorbic acid and dehydroascorbic acid. A further electron loss leads from the ascorbyl radical to dehydroascorbate. Dehydroascorbate still has low reducing abilities and can be oxidized further; it is reduced enzymatically to regenerate ascorbic acid. Therefore, ascorbic acid and its oxidation product dehydroascorbic acid form a reversible redox couple with the ascorbate radical as important intermediate contributing to the antioxidant function [9, 10, 13, 14]. Due to the reducing properties, ascorbic acid serves as one primary defense in the aqueous milieu. This scavenging ability is important in the eye where radiation and oxidative stress demand higher protection. Ascorbic acid interacts with glutathione and α-tocopherol as an antioxidant defense line [9, 11]. Ascorbate cannot directly scavenge lipophilic radicals occurring in membranes but it reduces the tocopheroxyl radicals bound in the membrane in the lipid-aqueous phase transition. In aqueous milieus, ascorbate protects glutathione, another water-soluble antioxidant of the cell, on its own expense and vice versa [9, 11, 13, 14]. The interaction of ascorbic acid with α-tocopherol and glutathione depicts a potent defense line in ocular tissue, protecting lens proteins and retina tissue of photooxidative damage.

Humans are unable to synthesize ascorbic acid and it is therefore actively absorbed from the diet by sodium-dependent transport systems in the intestine. The ascorbic acid concentrations of single tissues is tightly controlled and achieved with active cellular transporters for ascorbic acid and dehydroascorbic acid. The normal concentration of ascorbic acid in plasma is about 8–14 mg/l. As ocular tissue accumulates ascorbic acid, the concentrations are 100 times higher in the retina, 20-fold increased in the aqueous humor and in the lens the level is >10-fold the level of plasma [8, 9, 15]. Supplementation with ascorbic acid leads to a rise in the ascorbate levels in ocular tissue [8, 9]. The influence of ascorbic acid on the prevention of cataract and AMD is not yet confirmed. Epidemiological studies give inconsistent results. There are several studies concerned with nutritional supplement and cataract development, but due to their retrospective character they are prone to be biased prior to the diagnosis of cataract [16–18]. There are eight studies considering ascorbate intake and cataract prevalence. Four of these studies found a decreased prevalence for cataract when plasma ascorbic acid levels were high, some of these findings were not significant after adjustment for sex and age [8, 16, 18]. The other three studies found no association between ascorbic acid intake and plasma concentrations and prevalence of cataract. The last study showed an increased

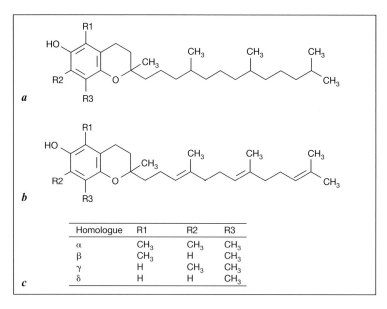

Homologue	R1	R2	R3
α	CH₃	CH₃	CH₃
β	CH₃	H	CH₃
γ	H	CH₃	CH₃
δ	H	H	CH₃

Fig. 2. Structures of tocopherol (**a**) and tocotrienol (**b**) and pattern of substitution of the homologues (**c**).

risk for cataract and high ascorbate plasma levels [16, 18]. There are reports indicating correlation between ascorbic acid intake and retinal levels, but for the progression of AMD increasing levels of carotenoids are of more importance than of ascorbate which had no influence at all on the progression of AMD [8, 16].

Tocopherols (Vitamin E)

Vitamin E is the most important lipid-soluble antioxidant in humans. It is a scavenger of peroxyl radicals and therefore inhibits the chain reactions in lipid peroxidation. The generic term 'vitamin E' is primarily a nutritional term and describes the eight tocopherol (α, β, γ, δ; fig. 2a) and tocotrienol (α, β, γ, δ; fig. 2b) homologues which exhibit vitamin E activity. Tocopherols are derivatives of the 2-methyl-6-chromanol with a phytyl side chain attached at position C-2; the α, β, γ and δ forms differing in the numbers and positions of their methyl groups at the ring (fig. 2c). The tocotrienols vary from the corresponding tocopherols in the isoprenoid side chain which is unsaturated at C-3′, C-7′ and C-11′. As α-tocopherol is the most abundant and active form in vivo, the term vitamin E refers now to α-tocopherol in the literature [3, 19, 20]. The methyl groups at the aromatic ring are crucial for the biological activity, the order of activity is: $\alpha > \beta > \gamma > \delta$. Furthermore, the hydroxyl group is important for the antioxidant activity of the tocopherols as it donates a hydrogen atom to free radicals [20]. The chain-braking properties are the result of the much faster reaction of tocopherols and tocotrienols with lipid peroxyl radicals than the reaction of these radicals with adjacent fatty acids and/or membrane proteins. In addition, tocopherols also react with singlet oxygen and superoxide radicals which contribute to their protective properties [3]. The arising tocopherol radicals can undergo different fates. They can react either with other tocopheryl radicals to give dimers or with other peroxyl radicals to give stable products [3, 19]. It is known that α-tocopheryl radicals are oxidized to α-tocopherylquinone which is metabolized and excreted in the urine [3]. The tocopheryl radicals can be recycled by reductants such as ubiquinol, polyphenolic compounds or ascorbic acid. The regeneration of the tocopherols with ascorbic acid is the major pathway for tocopheryl radicals in vivo and explains the synergistic antioxidant effects of these two antioxidants observed in vitro [3, 19, 20].

Vitamin E is absorbed by passive diffusion in the small intestine after its solubilization in micelles formed from fatty acids and bile acids. There are no differences in the absorption of the diverse homologues [20]. Vitamin E is transported in the blood within the lipoproteins and the normal plasma concentration in human ranges from 15 to 40 μM [21]. Supplementation of vitamin E results in an increase of the plasma level of about 2- to 3-fold [20]. The concentrations of α-tocopherol in the lens are about 1,573–2,550 ng/g wet weight and 257–501 ng/g wet weight for γ-tocopherol [22, 23]. It maintains the reduced glutathione levels in the lens and aqueous humor by enhancing the glutathione recycling [24, 25]. Supplementation of vitamin E does not enhance the concentration in the lens [8]. The rod outer segments and RPE contain high quantities of vitamin E. These tissues are sensitive to changing plasma levels of vitamin E [5]. Three studies correlated significantly high plasma α-tocopherol levels ($>20 \mu M$) with a lesser prevalence of cataract, this could not be confirmed by a fourth study [8, 18, 24, 25]. High dietary intake of vitamin E was not correlated with decreasing risks for cataract but dietary intake of vitamin E is hard to estimate as the use of ready-made foods and diverse brands of oil containing varying concentrations and compositions of tocopherols [5, 24]. There is no clear conclusion about pure vitamin E supplements; only one study found a significant lower prevalence for cataract whereas several other studies reported either non-significant inverse associations or no effects at all [18, 24, 25]. Combined supplements of multiple vitamins and/or minerals reduced the risk for cataract in several studies [8, 25]. As the retinal tissue is more sensitive to changes of vitamin E levels than the lens, in vitro studies suggest that

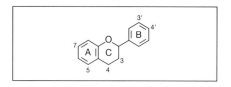

Fig. 3. Structure of flavan.

dietary intake of vitamin E or supplements have more influence in AMD than in cataract [5, 26]. There are only two studies (BDES and EDCC study) on this subject, the BDES study reporting a significant reduced risk for large macular drusen with high vitamin E intake. This significance was lost after adjustment for total vitamin E intake (dietary and supplement) [5, 8]. The other study found no significant association between dietary intake of vitamin E, with or without supplements, and AMD [5, 8]. High plasma levels of α-tocopherol were correlated significantly positive with a reduction of AMD in three studies, whereas two other studies found no significant associations [5, 8].

Flavonoids

Flavonoids belong to the secondary metabolites in plants stored as gluco-sides in the vacuole. The basic structure of flavonoids is the flavan structure (fig. 3). Precursors for the flavonoid synthesis are three malonyl-CoA (ring A and C) and one 4-cumaroyl-CoA (ring B) [4]. Flavonoids exhibit several posi-tive health aspects: they are anticancerogenic, antimutagenic, antiviral, antiox-idant, immune-stimulating and estrogen-active; they inhibit lipid peroxidation, low-densitiy lipoprotein (LDL) oxidation and chelate transition metals [27–29].

The maximal radical scavenging and/or antioxidant properties are given by [30, 31]: (a) the dihydroxyl groups at position 3′ and 4′ in ring B; (b) the 2,3-double bond in combination with the 4-oxo group in ring C; (c) the hydroxyl group at position 3 in ring C, and (d) the hydroxyl groups at position 5 and 7 in ring A.

After the reaction with radicals, the arising aroxyl radicals are stable enough not to undergo further chain reactions. The aroxyl radicals disproportionate on the one hand back to the parental flavonoid and otherwise to a quinoid structure [32]. Protective redox systems involving ascorbate and vitamin E can be extended with flavonoids interacting in a 'cascade' thus including lipophilic systems within this reaction. This was reported to work either synergistically or additive or show-ing a vitamin E or vitamin E and C sparing effect, respectively [33–36].

The knowledge about the bioavailability, uptake and metabolism is impor-tant to judge their pharmacological significance. Flavonoids occur in the diet as aglycones and glucosides. It is known that the aglycones as well as the glucosides pass the acidic conditions in the stomach unaltered [37]. They are either absorbed passively via diffusion as aglycones or actively via the sodium-dependent glucose/galactose transporters in the small intestine. Lactase/phlorizin hydrolase and cytosolic β-glucosidase in the brush border of the small intestine are able to hydrolyze the glucosides and enhance the concentration of the aglycones for free diffusion. The absorbed flavonoids reach the liver to undergo further metabolism. Glucosides are hydrolyzed from the liver and all existing aglycones are converted by enzymes of the detoxification metabolism into glucuronides, sulfates and/or methylates which are the circulating forms in blood/plasma besides ring fission products derived from colon metabolism [37–40]. Detected physiological concentrations of quercetin are 0.5 and 0.1 μM for isorhamnetin [39]. Numerous reports describe effects on eye diseases on the basis of plant extracts, mainly containing flavonoids: Extracts from *Primula macrophylla* containing 3′,4′-dihydroxychalcone and 3′-methoxyflavone are in use for several eye diseases in Pakistan and Afghanistan [41]; flavone glucosides (tetrahydroxy-trimethoxyflavone oligoglucosides) from the heart wood of *Pongamina pinnata* from India are used for healing eye diseases [42]; anti-cataractic properties on the basis of the inhibition of aldose reductase as an initiating enzyme of photo-oxidative and degenerative reactions in the lens are reported for flavonoids (acacetin, apigenin, luteolin, linarin) from '*Buddleja* Flos' or *Chrysanthemum boreale* [43, 44]. Furthermore, aqueous flavonoids from Propolis, the 'bee glue' used from bees to coat their hives, cure eye infections due to anti-inflammatory and antiviral effects [45]. An overview on natural therapies on ocular disorder is presented by Head [46]. The conclusion is that increased circulation to the optic nerve and antioxidant functions help to prevent and potentially to cure cataracts and glaucoma. Unfortunately, there are no reports or studies on the possible concentrations of flavonoids in the various tissues of the eye after either supplementation or treatment.

Synergistic Effects

The above-mentioned regeneration cycles for tocopherol, ascorbate and flavonoids were intensively investigated in our laboratory. At diene conjugation in the copper-induced LDL oxidation, a widely accepted assay for studies concerning lipid peroxidation occurring in vivo in LDL particles at the progress of atherogenesis, cooperative effects of α-tocopherol and ubiquinol were determined. Both compounds are consumed during diene conjugation in a clear pecking order: α-tocopherol disappears as soon as approximately 85% of ubiquinol is consumed and diene conjugation sets on with the complete consumption of

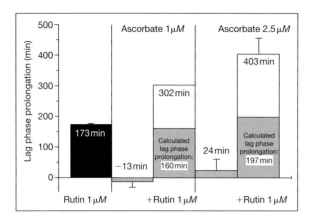

Fig. 4. Synergistic effects of ascorbic acid and rutin in copper-induced LDL oxidation.

α-tocopherol [47]. Synergistic implications of ascorbic acid and rutin, a naturally occurring flavonoid, were determined in LDL oxidation (fig. 4). Rutin is able to prolong the onset of diene conjugation (= lag phase prolongation), whereas ascorbic acid has either pro-oxidative ($1\,\mu M$) or a slight protective effect ($2.5\,\mu M$), respectively (fig. 4). Addition of rutin and ascorbic acid led to a synergistic lag time prolongation, indicating interactions between these two antioxidants which enhances their single antioxidant properties and prevents pro-oxidative effects [48]. LDL particles can be loaded with lipophilic antioxidants to investigate their impact on LDL oxidation and to study interactions between the loaded lipophilic compound and hydrophilic samples added to the assay. In the case of lycopene- or lutein-loaded LDL, no significant prolongation was determined (fig. 5), but the addition of rutin led to a synergistic prolongation of the lag time (fig. 5), indicating interactions between the lipophilic antioxidant in the LDL particle and the hydrophilic compound in the surrounding aqueous milieu of the assay [49]. These results support the findings of several eye studies on AMD and cataract prevention which supplemented with multivitamin preparations. The results of the studies indicate that a reasonable composition of multivitamins possibly enriched with minerals provides a reduction for the incident of cataract and AMD [8, 16, 18, 25, 26].

Conclusion

The presented data show clearly that the antioxidant status of the eye plays a crucial role in pathogenesis of AMD and cataract, but other factors such as genetics, way of life or environmental influences need to be considered as well.

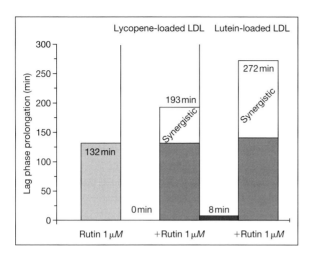

Fig. 5. Synergistic effects of lycopene- or lutein-loaded LDL with rutin in copper-induced LDL oxidation.

The influence of flavonoids has to be studied intensively to create a valid data basis. Furthermore, the data for vitamin C and E are inconsistent at the moment and need onward investigations. At least the combination of single antioxidants such as vitamin C and E and flavonoids are a possible field for future studies, but carotenoids need to be included as they contribute extensively to the antioxidant status in the eye.

References

1   Elstner EF: Der Sauerstoff: Biochemie, Biologie, Medizin. Mannheim, BI Wissenschaftsverlag, 1990.
2   Elstner EF: Sauerstoffabhängige Erkrankungen und Therapien. Mannheim, BI Wissenschaftsverlag, 1993.
3   Halliwell B, Gutteridge JMC: Free Radicals in Biology and Medicine. Oxford, Oxford University Press, 1999.
4   Karlson P, Doenecke D, Koolman J: Kurzes Lehrbuch der Biochemie für Mediziner und Naturwissenschaftler. Stuttgart, Thieme, 1994.
5   Beatty S, Koh HH, Henson D, Boulton M: The role of oxidative stress in the pathogenesis of age-related macular degeneration. Surv Ophthalmol 2000;45:115–134.
6   Noske UM, Schmidt-Erfurth U, Meyer C, Diddens H: Lipidmetabolismus im retinalen Pigmentepithel. Ophthalmologe 1998;95:814–819.
7   Keilhauer CN, Weber BHF: Die altersabhängige Makuladegeneration – eine häufig multifaktorielle Erkrankung des höheren Alters. BIOspektrum 2003;1:19–22.
8   Brown NAP, Bron AJ, Harding JJ, Dewar HM: Nutrition supplements and the eye. Eye 1998;12:127–133.
9   Tsao CS: An overview of ascorbic acid chemistry and biochemistry; in Packer L, Fuchs J (eds): Vitamin C in Health and Disease. New York, Dekker, 1997, pp 25–58.

10   Bors W, Buettner GR: The vitamin C radical and its reactions; in Packer L, Fuchs J (eds): Vitamin C in Health and Disease. New York, Dekker, 1997, pp 75–94.

11   Stocker R, Frei B: Endogenous antioxidant defences in human blood plasma; in Oxidative Stress, Oxidants and Antioxidants. London, Academic Press, 1991, pp 215–243.

12   Duthie GG: Determination of activity of antioxidants in human subjects. Proc Nutr Soc 1999;58:1015–1024.

13   Packer L: Vitamin C and redox cycling antioxidants; in Packer L, Fuchs J (eds): Vitamin C in Health and Disease. New York, Dekker, 1997, pp 95–105.

14   Basu TK: Potential role of antioxidant vitamins; in Basu TK, Temple NJ, Garg ML (eds): Antioxidants in Human Health and Disease. New York, CAB International, 1999, pp 15–26.

15   Varma SC, Devamanoharan PS, Ali AH: Oxygen radicals in the pathogenesis of cataracts – Possibilities for therapeutic intervention; in Taylor A (ed): Nutritional and Environmental Influences on the Eye. Boca Raton, CRC Press, 1999, pp 5–24.

16   Taylor A, Dorey DK, Nowell T: Oxidative stress and ascorbate in relation to risk for cataract and age-related maculopathy; in Packer L, Fuchs J (eds): Vitamin C in Health and Disease. New York, Dekker, 1997, pp 231–263.

17   Mares-Perlman JA, Klein R: Diet and age-related macular degeneration; in Taylor A (ed): Nutritional and Environmental Influences on the Eye. Boca Raton, CRC Press, 1999, pp 181–214.

18   Taylor A: Nutritional and environmental influences on risk for cataract; in Taylor A (ed): Nutritional and Environmental Influences on the Eye. Boca Raton, CRC Press, 1999, pp 53–94.

19   Niki E, Noguchi N: Dynamics of antioxidant action of vitamin E. Acc Chem Res 2004;37:45–51.

20   Bramley PM, Elmadfa I, Kafatos A, Kelly FJ, Manios Y, Roxborough HE, Schuch W, Sheehy PJA, Wagner KH: Vitamin E – Review. J Sci Food Agric 2000;80:913–938.

21   Thews G, Mutschler E, Vaupel P: Anatomie, Physiologie, Pathophysiologie des Menschen. Stuttgart, Wissenschaftliche Verlags-Gesellschaft, 1999.

22   Yeum KJ, Taylor A, Tang G, Russell RM: Measurement of carotenoids, retinoids, and tocopherols in human lenses. Invest Ophthalmol Vis Sci 1995;36:2756–2761.

23   Yeum KJ, Shang F, Schalch W, Russell RM, Taylor A: Rat-soluble nutrient concentrations in different layers of human cataractous lens. Curr Eye Res 1999;19:502–505.

24   Wu SY, Leske MC: Antioxidants and cataract formation: A summary review. Int Ophthalmol Clin 2000;40:71–81.

25   Taylor A, Nowell T: Oxidative stress and antioxidant function in relation to risk for cataract; in Advances in Pharmacology. London, Academic Press, 1997, pp 515–536.

26   Richer S: Antioxidants and the eye. Int Ophthalmol Clin 2000;40:1–16.

27   Rice-Evans CA, Miller NJ, Bolwell GP, Bramley PM, Pridham JB: The relative antioxidant activities of plant-derived polyphenolic flavonoids. Free Radic Res 1995;22:375–383.

28   Bors W, Michel C, Stettmaier K: Antioxidant effects of flavonoids. BioFactors 1997;6:399–402.

29   Hider RC, Liu ZD, Khodr HH: Metal chelation of polyphenols; in Packer L (ed): Methods in Enzymology. London, Academic Press, 2001, vol 335, pp 190–203.

30   Bors W, Heller W, Michel C, Saran M: Flavonoids as antioxidants: Determination of radical-scavenging efficiencies; in Packer L, Glazer AN (eds): Methods in Enzymology. London, Academic Press, 1990, vol 186, pp 343–355.

31   Heijnen CGM, Haenen GRM, Oostveen M, Stalpers EM, Bast A: Protection of flavonoids against lipid peroxidation: The structure-activity relationship revised. Free Radic Res 2002;36:575–581.

32   Bors W, Michel C, Stettmaier K: Structure-activity relationships governing antioxidant capacities of plant polyphenols; in Packer L (ed): Methods in Enzymology. London, Academic Press, 2001, vol 335, pp 166–180.

33   Lotito SB, Fraga CG: (+)-Catechin as antioxidant: Mechanisms preventing human plasma oxidation and activity in red wines. BioFactors 1999;10:125–130.

34   Liao KL, Yin MC: Individual and combined antioxidant effects of seven phenolic agents in human erythrocyte membrane ghosts and phosphatidylcholine liposome systems: Importance of the partition coefficient. J Agric Food Chem 2000;48:2266–2270.

35   Vasiljeva OV, Lyubitsky OB, Klebanov GI, Vladimirow YA: The effect of combined action of flavonoids, ascorbate and alpha-tocopherol on peroxidation of phospholipid liposomes, induced by the Fe^{2+} ions. Biologich Membr 2000;17:42–49.

36   Miller NJ, Ruiz-Larrea MB: Flavonoids and other plant phenols in the diet: Their significance as antioxidants. J Nutr Environ Med 2002;12:39–51.

37   Gee JM, Dupont SM, Rhodes MJC, Johnson IT: Quercetin glucosides interact with the intestinal glucose transport pathway. Free Radic Biol Med 1998;25:19–25.

38   Day AJ, Bao Y, Morgan MRA, Williamson G: Conjugation position of quercetin glucuronides and effect on biological activity. Free Radic Biol Med 2000;29:1234–1243.

39   Day AJ, Mellon F, Barron D, Sarrazin G, Morgan MRA, Williamson G: Human metabolism of dietary flavonoids: Identification of plasma metabolites of quercetin. Free Radic Res 2000;35:941–952.

40   Wittig J, Herderich M, Graefe EU, Veit M: Identification of quercetin glucuronides in human plasma by high-performance liquid chromatography-tandem mass spectrometry. J Chromatogr B 2001;753:237–243.

41   Ahmad VU, Shah MG, Mushtaq-Noorwala, Mohammad FV: Isolation of 3,3'-dihydroxychalcone from *Primula macrophylla*. J Nat Prod 1992;55:956–958.

42   Babita-Agrawal H, Singh J, Agrawal B: Two new flavone glycosides from (heartwood) of *Pongamia pinnata*. Int J Pharmacognosy 1993;31:305–310.

43   Matsuda H, Cai H, Kubo M, Tosa H, Iinuma M: Study on anti-cataract drugs from natural sources. II. Effects of *Buddlejae* Flos on in vitro aldose reductase activity. Biol Pharm Bull 1995;18:463–466.

44   Shin KH, Kang SS, Seo EA, Shin SW: Isolation of aldose reductase inhibitors from the flowers of *Chrysanthemum boreale*. Arch Pharm Res 1995;18:65–68.

45   Crisan I, Petica M, Mutiu A: Some morphopathological aspects of the experimental eye infection with herpes simplex virus type 1 in rabbits, followed by a treatment with aqueous flavonoids solution obtained from Propolis. Apicata 1996;31:72–80.

46   Head KA: Natural therapies for ocular disorders. 2. Cataract and glaucoma. Altern Med Rev 2001;6:141–166.

47   Schneider D, Elstner EF: Coenzyme Q_{10}, vitamin E, and dihydrothioctic acid cooperatively prevent diene conjugation in isolated low-density lipoprotein. Antioxid Redox Signal 2000;2:327–333.

48   Milde J, Elstner EF, Grassmann J: Synergistic inhibition of low-density lipoprotein oxidation by rutin, γ-terpinene, and ascorbic acid. Phytomedicine 2004;11:105–113.

49   Milde J: Kooperative Wirkung pflanzlicher Antioxidantien in pathologisch relevanten Arteriosklerose- und Arthritismodellen; thesis, Freising, 2004.

K.M. Janisch
TUM Weihenstephan, Center of Life and Food Sciences
Lehrstuhl für Phytopathologie
Am Hochanger 2, DE–85350 Freising (Germany)
E-Mail k.janisch@agrar.tu–muenchen.de

Augustin A (ed): Nutrition and the Eye.
Dev Ophthalmol. Basel, Karger, 2005, vol 38, pp 70–88

....

Macular Carotenoids: Lutein and Zeaxanthin

W. Stahl

Heinrich Heine University Düsseldorf, Institute of Biochemistry and
Molecular Biology I, Düsseldorf, Germany

Abstract

The yellow color of the macula lutea is due to the presence of the carotenoid pigments
lutein and zeaxanthin. In contrast to human blood and tissues, no other major carotenoids
including β-carotene or lycopene are found in this tissue. The macular carotenoids are
suggested to play a role in the protection of the retina against light-induced damage.
Epidemiological studies provide some evidence that an increased consumption of lutein and
zeaxanthin with the diet is associated with a lowered risk for age-related macular degeneration, a disease with increasing incidence in the elderly. Protecting ocular tissue against photooxidative damage carotenoids may act in two ways: first as filters for damaging blue light,
and second as antioxidants quenching excited triplet state molecules or singlet molecular
oxygen and scavenge further reactive oxygen species like lipid peroxides or the superoxide
radical anion.

Carotenoids comprise a class of natural lipophilic pigments which are
found in plants, algae, bacteria, yeasts, and molds [1]. They are responsible for
many of the yellow, orange and red hues of fruits and flowers. Chlorophyll
masks the carotenoids in green leaves but in autumn, as the chlorophyll levels
decline, the color of the carotenoids becomes visible and produces the yellows
and reds of autumn foliage. As accessory pigments, carotenoids participate in
photosynthetic processes and are involved in mechanisms of photoprotection in
higher plants, dissipating excess light energy through the xanthophyll cycle,
with the formation of zeaxanthin from violaxanthin [2]. Carotenoids can also
be found in many animal species, and are important colorants in birds, insects,
fish, and crustacean, although animals are not capable of synthesizing
carotenoids de novo and depend on dietary supply. About 600 different

carotenoids have been characterized and new ones continue to be identified [3]. Among the huge variety of structurally different carotenoids about 50 occur in the human diet with β-carotene being the most prominent [4, 5]. Epidemiological studies clearly show that the consumption of a diet rich in fruit and vegetables is correlated with a lower risk for a number of diseases including some types of cancer, as well as cardiovascular, neurodegenerative and ophthalmological disorders [6]. Among the dietary components, micronutrients have been suggested to be involved in the protection against such age-related diseases [7]. Carotenoids are dietary constituents, provided in high amounts by fruit and vegetables, and they likely play a role in disease prevention [8]. β-Carotene and some other members of the carotenoid family are so-called provitamin A compounds. After absorption, they are cleaved by specific enzymes and significantly contribute to human vitamin A supply which is one of the most important biological features of carotenoids. Vitamin A is essential for vision, growth and development. However, carotenoids also reveal other biological properties apparently contributing to health and to the prevention of diseases [9, 10]. Most of the carotenoids, including the major dietary non-provitamin A compounds such as lutein, zeaxanthin, and lycopene, are very efficient antioxidants, provide photoprotection, and trigger cellular communication.

Within the last decade possible health effects of the carotenoids lutein and zeaxanthin have attracted attention, and levels of adequate supply were discussed with respect to beneficial effects on ocular health [11, 12]. There is increasing evidence from epidemiological studies that an increased intake of the macular pigments lutein and zeaxanthin is inversely associated with the risk for age-related macular degeneration (AMD), a disease which affects the elderly and is a major cause of irreversible blindness in Western countries. Although effects of lutein and zeaxanthin in the prevention of AMD remain to be proven in intervention studies, they exhibit physicochemical and biochemical properties which make them suitable compounds for photoprotection of the retina.

It should be noted that various synthetic carotenoids and extracts from carotenoid-rich plants are also used as food colorants, additives to animal feeds, nutritional supplements and for cosmetical and pharmaceutical purposes. In many multivitamin formulas, single carotenoids or carotenoid mixtures are included.

Structure and Biosynthesis

The chemical structures of lutein, zeaxanthin and β-carotene are presented in figure 1. All members of the carotenoid family are tetra-terpenoids composed of a central carbon chain with conjugated double bonds carrying different linear

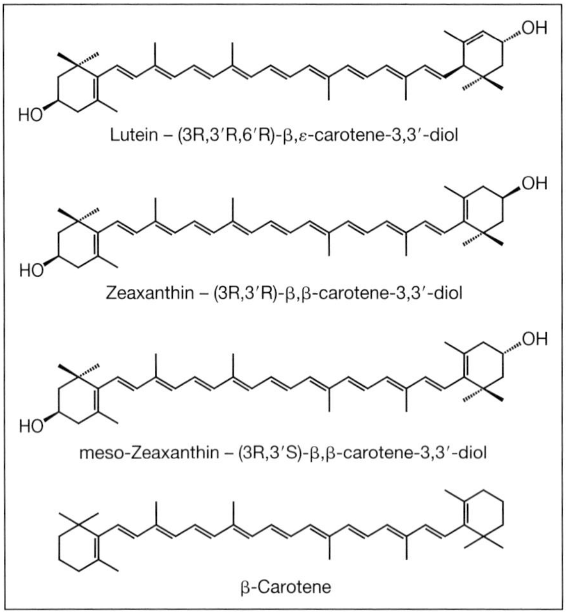

Fig. 1. Structures of selected carotenoids.

or cyclic substituents. Based on their composition, carotenoids are divided into two subgroups: the oxygen free carotenes and the xanthophylls (oxo-carotenoids) which contain at least one oxygen atom in their structure. Lutein and zeaxanthin are substituted with two hydroxyl groups at the 3 and 3′-position of the ionone rings and can be formally assigned as dihydroxy derivatives of α- and β-carotene, respectively. Due to the presence of hydroxyl groups they are more polar than carotenes. In both compounds, nine carbon-carbon double bonds of the polyene backbone are fully conjugated, whereas the double bonds in the ring are only partially in conjugation. The pattern of conjugated double bonds determines the light-absorbing properties and influences the antioxidant activity of carotenoids. The absorption maximum of lutein is around 445 nm, the one of zeaxanthin 450 nm; the molar extinction coefficients (ε_{mol}) at these wavelengths are in the range of 140,000–145,000 cm^{-1}mol^{-1}. Thus, both carotenoids are efficiently absorbing blue light [13]. Because of the presence of chiralic centers, lutein and zeaxanthin may occur in several stereoisomeric forms, three in the case of zeaxanthin and eight for lutein. In plants only one major stereoisomer of lutein, [(3R,3′R,6′R)-β,ε-carotene-3,3′-diol], and one of zeaxanthin, [(3R,3′R)-β, β-carotene-3,3′-diol] is found because of stereospecific biosynthesis. Consequently, these isomers are also dominant in our diet and the ones preferentially provided to the organism. According to the number of double bonds, several *cis/trans* (E/Z) configurations are possible for a given molecule. In homogenous solutions, carotenoids tend to isomerize and form a mixture of mono- and poly-*cis* isomers in addition to the all-*trans* form. When incorporated into the food matrix the compounds are more resistant towards isomerization. Generally, the all-*trans* form is predominant in nature but several *cis* isomers of carotenoids are present in blood and tissues [10, 14]. Absorption spectra of carotenoid *cis* isomers exhibit an additional absorption band at a characteristic position about 120 nm below the wavelength of the absorption maximum [13]. The intensity of the so-called '*cis* peak' depends on the position of the *cis* bond and is most intensive when the central carbon-carbon double bond of the molecule is in *cis* configuration.

The key molecule in carotenoid biosynthesis is isopentenyl diphosphate used to build up 'step by step' the carotenoid phytoene [15]. Several enzymatic dehydrogenation reactions lead to the acyclic carotenoid lycopene. α-Carotene, β-carotene, lutein, zeaxanthin and an array of other carotenoids are synthesized via subsequent cyclization, dehydrogenation, and oxidation reactions. A great number of enzymes involved in carotenoid biosynthesis have been identified and their DNA sequence has been determined. Methods of modern gene technology provide the tools to modulate carotenoid biosynthesis in plants and microorganisms and to change their carotenoid pattern [15]. Single carotenoids or groups of carotenoids may be enriched by directed synthesis and genetically modified plants, algae or microorganisms may be used as suitable sources for selected carotenoids.

Occurrence in Food

Lutein and zeaxanthin are among the most abundant carotenoids in our diet and mainly provided by fruit and vegetables. Carotenoid analyses have improved in the last decade, however, before the introduction of suitable stationary phases it was difficult to separate and quantify lutein and zeaxanthin with standard chromatographic methods. Therefore, their levels in carotenoid databases for foods are often reported as the sum of both compounds (table 1). In the US National Health and Education Survey, the intake provided with a Western diet has been estimated to be 1.3–3.0 mg/day of lutein and zeaxanthin combined [16]. In the German National Food Consumption Survey an average lutein intake of 1.9 mg/day was calculated, slightly increasing with age [17]. Lutein contributed about 35% to the total carotenoid consumption. Similar data have been reported for other countries. The Food Habits of Canadians Study provided intake data for lutein (adults aged 18–65) as being 1.4 mg/day [18].

Table 1. Dietary sources of lutein and zeaxanthin

Dietary source	Lutein + zeaxanthin μg/100 g edible portion
Beans	640–700
Broccoli	830–2,450
Brussels sprouts	1,290–1,590
Cabbage	310
Corn	880–1,800
Kale	15,800–39,550
Lettuce	350–2,640
Oranges	190
Orange juice	40–140
Peaches	30–60
Peas (green)	1,350
Spinach	7,040–11,940
Squash	40–2,130
Tangerines	240
Tangerine juice	170
Tomatoes	40–130
Tomato products	0–170

However, the studies also show that there is a broad range of intakes which apparently depends on individual preferences for specific foods. In the German study, green leafy salads and spinach have been identified as the major sources of lutein and zeaxanthin providing about 50% of the total supply. Lettuce, spinach, corn, broccoli and oranges were identified as the foods contributing most of the lutein to the Canadian diet. Further important sources are egg yolks, other green vegetables and fruits such as green beans, green peas, brussels sprouts, cabbage, kale, or peaches and tangerines. Also, fruit juices contribute to the intake of xanthophylls. In most plants the content of lutein by far exceeds that of zeaxanthin and ratios of 7:1 to 4:1 have been reported. Corn contains quite high amounts of zeaxanthin and it has been suggested that it is an important source for this specific carotenoid. Also, some potato varieties contain considerable amounts of lutein and zeaxanthin with the latter dominating in some of the strains [19].

Lutein and zeaxanthin are further found in algae, other microorganisms and the petals of many yellow flowers which are used as natural sources of these carotenoids. The hydroxy groups of both compounds can be esterified with various naturally occurring fatty acids [20, 21]. Thus, mono- and diesters of lutein and zeaxanthin also occur in plants. Lutein and zeaxanthin dipalmitates,

Table 2. Carotenoid serum or plasma levels (nmol/ml)

Ref.	Subjects (n)	Lutein + zeaxanthin	β-Cryptoxanthin	Lycopene	α-Carotene	β-Carotene	Sum of carotenoids
84	111	0.26	0.32	0.39	0.05	0.22	1.24
85	33	0.35	0.17	0.26	0.08	0.25	1.11
86	3,480	0.36	0.22	0.40	0.08	0.34	1.40
29	54	0.18	0.25	0.22	nd	0.38	1.03
87	57	0.49	0.24	1.06	0.09	0.58	2.46
88	98	0.46	0.17	0.58	0.11	0.34	1.66
22	220	0.27	0.17	0.43	0.08	0.32	1.27
22	180	0.25	0.12	0.43	0.06	0.21	1.07
Mean ± SD		0.33 ± 0.11	0.21 ± 0.06	0.47± 0.26	0.08 ± 0.02	0.33 ± 0.12	1.41 ± 0.47

nd = Not determined.

dimyristates, several mixed esters as well as monomyristates and further monoesters have been identified in the petals of the marigold flower (*Tagetes erecta*). Depending on sources and processing, nutritional supplements may contain lutein esters, with much smaller amounts of zeaxanthin esters, and/or free lutein and zeaxanthin.

Absorption, Distribution and Metabolism

Generally, the carotenoid pattern in human plasma is determined by the variety of fruit and vegetables ingested within the diet [22]. Table 2 lists the blood levels of the major carotenoids reported in several studies. In the different studies, average lutein plus zeaxanthin concentrations were in the range of 0.18–0.49 nmol/ml. The ratio between lutein and zeaxanthin is usually between 3:1 and 5:1 [23]. Both compounds are also found, together with other members of the carotenoid family, in human tissues including liver, kidney, lung, and skin [24, 25]. In skin, small amounts of carotenol fatty acid esters were identified, among them are lineolate, palmitate, oleate, myristate, and stearate mono- and diesters of lutein, zeaxanthin, 2′,3′-anhydrolutein, α-cryptoxanthin and β-cryptoxanthin [26]. It has been suggested that carotenol esters in human skin may be formed by re-esterification of xanthophylls following absorption.

Carotenoid uptake from the diet follows the pathway of lipophilic nutrients. After digestion of food, carotenoids are incorporated into micelles formed from dietary lipids and bile acids, which facilitate absorption into the intestinal mucosal cell. Further, carotenoids are incorporated into chylomicrons which are released into the lymphatic system. In the blood, carotenoids appear initially in the chylomicron and VLDL fraction, whereas the levels in other lipoproteins such as LDL and HDL rise at later time points with peak levels at 24–48 h. The major vehicle of hydrocarbon carotenoids such as lycopene and β-carotene is the LDL, whereas the polar xanthophylls are more equally distributed between LDL and HDL.

The bioavailability of carotenoids is influenced by several factors and may vary within a wide range. Dietary fat, consumed together with the carotenoids, improves the absorption. Lutein plasma responses were higher when lutein esters were consumed together with a high-fat spread compared to coingestion with a low-fat meal [27]. On the contrary, bioavailability is decreased when certain types of dietary fiber are coingested. Depending on the type of fiber, plasma levels of lutein and other carotenoids, calculated as area under the plasma response curve ($AUC_{0-24 h}$), were diminished by 40–75% [28]. When carotenoids are consumed together with non-absorbable fat replacers like sucrose polyesters, absorption is impaired [29]; xanthophylls are less affected than carotenes.

Processing of carotenoid-containing products usually improves bioavailability. Disruption of cellular structures and release of the compounds from the food matrix as well as improved accessibility to lipophilic subcellular compartments have been suggested to play a role [30]. When carotenoids were supplied with a diet rich in vegetables, the bioavailability of lutein was higher than that of β-carotene [31]. After ingestion of a supplement rich in β-carotene which contained small amounts of lutein and zeaxanthin a preferential increase of the xanthophylls in the chylomicron fraction of lipoproteins was observed [32].

In human and animal studies, interactions between carotenoids during absorption have been described [33]. Lutein exhibited an inhibitory effect on β-carotene uptake which was most pronounced when lutein was the predominant carotenoid in the source. β-Carotene cleavage, however, was not affected [34]. Competition between single carotenoids for incorporation into micelles and exchange of the compounds between lipoproteins apparently play a role, and may explain the different blood responses.

In dietary supplements nowadays on the market and containing lutein, the carotenoid may be present in two different forms – either esterified with different fatty acids or unesterified as so-called 'free' lutein. There is an ongoing debate regarding possible differences in the bioavailability of lutein from both forms. However, present data do not provide evidence that lutein from fatty acid esters is less bioavailable than 'free' lutein [35]. Other factors such as dissolution of the formulation appear to be more important and should be considered for the evaluation of xanthophyll bioavailability. After ingestion of a single dose of zeaxanthin dipalmitate from wolfberry or non-esterified zeaxanthin (equal amounts), somewhat higher AUC values were determined in the group supplemented with the ester [36]. No differences in plasma responses were found after application of β-cryptoxanthin esters from papaya and non-esterified β-cryptoxanthin [37]. Based on the present data it is suggested that ester hydrolysis is not a limiting step in lutein and zeaxanthin absorption and that the esters exhibit at least an equivalent bioavailability compared to the 'free' compounds.

Esters of carotenoids are usually not detectable in human blood. However, when esterified lutein was supplied for a longer period of time, free lutein concentrations increased significantly, and in addition small amounts of lutein esters were detectable in the serum of some of the volunteers [38]. However, the contribution of lutein esters to total lutein was less than 3%.

Present knowledge on lutein and zeaxanthin metabolism is scarce, but metabolism and conversion play a role in the formation of the specific carotenoid pattern of the macula lutea (see below). In studies supplementing lutein and zeaxanthin, metabolites of the compounds have been detected in serum but have been characterized only partially [16]. Several carotenoids appearing in human serum are suggested to be derived from both xanthophylls, either by dehydration or oxidation of the hydroxyl group, the latter yielding mono- and di-keto derivatives of the parent carotenoids [5].

The Macular Pigment

The identity of lutein and zeaxanthin as major carotenoids in the macular pigment of humans and other primates has been well established [39–43]. Both carotenoids have been identified by UV-Vis spectroscopy, mass spectrometry and by chromatographic methods comparing elution profiles with authentical standards. They are also dominating the carotenoid pattern of the entire retina but the concentration in the macula lutea is about 100 times higher than in the peripheral retina. It has been calculated that the levels of the macular pigments range from 0.3 to 1.3 mM [44], which is much higher than carotenoid concentrations found in any other tissue.

Limited information is available on the effect of carotenoid supplementation and macula pigment density. However, first studies show that upon supplementation with lutein the density of the macula pigment increases [45].

In addition to lutein and zeaxanthin, an oxidation product of lutein (3-hydroxy-β,ε-carotene-3′-one) was identified; some other oxo-carotenoids were also found but in minor amounts [42]. Several geometrical isomers of lutein and zeaxanthin, including 9-*cis*-lutein, 9′-*cis*-lutein, 13-*cis*-lutein, 13′-*cis*-lutein, 9-*cis*-zeaxanthin, and 13-*cis*-zeaxanthin were determined in human and monkey retinas. It is unknown if the *cis* isomers have specific roles compared the all-*trans* forms of the macular carotenoids.

Although only one stereoisomeric form of lutein, (3R,3′R,6′R)-β, ε-carotene-3,3′-diol, and of zeaxanthin, (3R,3′R)-β,β-carotene-3,3′-diol, is provided with the diet, considerable amounts of a second stereoisomer of zeaxanthin, (3R,3′S)-β,β-carotene-3,3′-diol, were identified in the central region of macula lutea applying chiral column chromatography [46]. The latter isomer has been assigned as meso-zeaxanthin and differs from 'natural' zeaxanthin by the configuration at the 3′-carbon atom. The UV-Vis spectra of zeaxanthin and meso-zeaxanthin are identical. Meso-zeaxanthin is not detectable in considerable amounts in other human tissues or blood [46, 47]. Thus, it has been suggested that the isomer is specifically formed in ocular tissues. Two different mechanisms have been discussed both assuming that meso-zeaxanthin is generated from the lutein stereoisomer (3R,3′R,6′R)-β,ε-carotene-3,3′-diol. One pathway proposes the intermediate formation of an oxidation product involving the allylic 3′-hydroxy group of lutein, the other suggests that the isomerization results from a direct migration of the isolated double bond in the ε-ring of lutein, preserving the configuration at the 3′-carbon atom [16, 42, 47].

Interestingly, lutein, zeaxanthin and meso-zeaxanthin levels and ratios vary in different regions of the macula lutea [16, 48]. Moving from excentric positions of the retina towards the fovea (center) carotenoid concentrations increase and the lutein to zeaxanthin ratio changes in favor of zeaxanthin. Lutein is the dominating carotenoid in the outer segments of the retina (lutein:zeaxanthin about 2:1), whereas zeaxanthin is the major carotenoid in the center and the ratio is reversed. In accordance with dietary supply, the amount of lutein in human serum exceeds that of zeaxanthin by the factor 3–4. Applying chiral chromatography, the distribution pattern of zeaxanthin isomers (3R,3′R)-β,β-carotene-3,3′-diol, and (3R,3′S)-β,β-carotene-3,3′-diol (meso-zeaxanthin) was analyzed. Both isomers are found in about equal amounts (ratio 1:1) in the center of the macula lutea, whereas the contribution of meso-zeaxanthin to total zeaxanthin is gradually decreasing moving to peripheral areas. The reasons for the enrichment of meso-zeaxanthin in the central area and the spatial distribution pattern of lutein and zeaxanthin are unknown.

It is generally accepted that xanthophyll-binding proteins with high affinity for the hydroxylated ionone rings are involved in the directed transport of macular carotenoids. However, detailed knowledge on the structures and properties of these proteins is lacking. It is also unknown whether the spatial and structural distribution of carotenoids within the retina is related to specific functions and, if so, what biochemical or physicochemical properties may underlie such functional aspects. There is some evidence that the orientation of xanthophylls in cellular membranes might play a crucial role.

Analysis of transverse retina sections shows that high levels of the carotenoids are present in the photoreceptor axons. Tubulin is found in abundance in the receptor axon layer of the fovea. There is evidence for specific xanthophyll-tubulin interactions and it has been suggested that the protein serves as a locus for the deposition of macular carotenoids [49]. Substantial amounts of lutein and zeaxanthin were also determined in rod outer segments [50]. Based on the localization of the carotenoids and their chemical properties it has been suggested that they protect photoreceptors by filtering blue light and provide additional protection by scavenging reactive oxygen species.

Oxidative Stress

The development and existence of an organism in the presence of oxygen is associated with the generation of reactive oxygen species (ROS), even under physiological conditions. ROS are responsible for the oxidative damage which may affect all types of biological molecules, including DNA, lipids, protein and carbohydrates [51–53]. Such damage is discussed as pathobiochemical mechanism playing a crucial role in the development of various diseases [54, 55]. Some of the most relevant ROS are: singlet oxygen ($^1O_2$), peroxyl radicals (ROO$^\cdot$), the superoxide anion radical ($O_2^{\cdot-}$), the nitric oxide radical (NO$^\cdot$), peroxynitrite (ONOO$^-$), and hydrogen peroxide (H_2O_2). ROS are either radicals (molecules that contain at least one unpaired electron) or reactive non-radical compcunds such as excited state molecules like $^1O_2$. These reactive intermediates are often summarized by the term oxidants or prooxidants.

The prooxidant load is counteracted by a diversity of antioxidant defense systems operative in biological systems which include antioxidant enzymes, low-molecular-weight antioxidants, trace elements and specific proteins. An antioxidant has been defined as 'any substance that, when present in low concentrations compared to that of an oxidizable substrate, significantly delays or inhibits the oxidation of that substrate' [51–53].

The major enzymes directly involved in the detoxification of ROS are superoxide dismutase, which is scavenging $O_2^{\cdot-}$, as well as catalase and glutathione peroxidases reducing hydrogen peroxide and organic hydroperoxides, respectively.

Several endogenous low-molecular-weight compounds are also involved in antioxidant defense, of which glutathione is the most prominent. Other endogenous compounds such as ubiquinol-10, urate, or bilirubin also contribute to antioxidant defense.

With the human diet, an array of different compounds possessing antioxidant activities are provided to the organism. The most prominent representatives of dietary antioxidants are carotenoids, ascorbate (vitamin C), tocopherols (vitamin E), and polyphenols. Most groups of dietary antioxidants comprise a number of structurally different compounds [54]. Taken together, the organism is protected against oxidative damage by a network of defense systems, which may act synergistically and fulfill specific tasks depending on their reactivity and distribution in tissues and at the subcellular level.

When ROS are produced in excessive amounts and not sufficiently detoxified, the steady-state balance between the prooxidant load and the antioxidant network may be disrupted. A disbalance in favor of the prooxidants potentially leading to damage has been defined as 'oxidative stress' [56].

Photooxidative Stress

Upon light exposure, a cascade of photo-induced reactions takes place in the exposed tissues [57, 58]. As primary event, light interacts with a suitable chromophore which mediates desired biological responses but may also initiate damaging reaction sequences. Photo-induced damage may directly affect the chromophore, e.g. formation of pyrimidine dimers in the DNA. However, the chromophor can also act as photosensitizer initiating subsequent photochemical reactions. Porphyrins, flavins, DNA bases, amino acids or lipofuscin have been shown to act as photosensitizers. Upon illumination, the photosensitizer is excited from the ground state to the first excited singlet state, followed by conversion to the triplet state via intersystem crossing. According to the post-excitational chemistry the excited state can react in two ways – either by type I or type II photooxidation reactions [58–60]. A type I mechanism involves hydrogen-atom abstraction or electron-transfer reactions between the excited state of the photosensitizer and a cellular substrate, yielding free radicals and radical ions. Such intermediates undergo further reaction sequences, leading to the deterioration of proteins, lipids or DNA and finally damage important cellular structures. As the primary reaction product of a type II photooxidation reaction, singlet molecular oxygen ($^1O_2$) is generated via energy transfer from the excited photosensitizer to ground state oxygen. $^1\Delta_g O_2$, the major form of electronically excited, singlet oxygen, is a dienophilic intermediate which reacts with polyunsaturated fatty acids, DNA bases, histidine, tyrosine or tryptophan residues forming cyclic or acyclic peroxides. Peroxides are unstable compounds and undergo further rearrangement, elimination or Fenton-type reactions, yielding modified biomolecules or free radicals. Again, radical-initiated reaction chains take place and impart damage to biological and cellular structures.

Lipofuscin, a heterogenous conglomerate of biomolecules and breakdown products, has been assigned as the 'aging pigment' and is found in various postmitotic cells. In the retinal pigment epithelium, lipofuscin is present as micrometer-sized spherical particles and an accumulation has been observed in the process of macular degeneration. The pigment is characterized by its yellow autofluorescence upon exposure to blue light and it has been suggested that fluorescent components of lipofuscin granules are at least in part responsible for phototoxic reactions [61]. The composition of lipofuscin is poorly defined and the fluorophores have only been partially characterized. However, it has been demonstrated that retinal lipofuscin is a photoinducible generator of ROS, and illumination of lipofuscin with visible light leads to extragranular lipid peroxidation, enzyme inactivation, and protein oxidation [62].

Carotenoids as Antioxidants

Carotenoids, including lutein and zeaxanthin, efficiently scavenge ROS which is considered to be an important biological task of these compounds. According to their unique structure they are efficient scavengers of singlet molecular oxygen and of peroxyl radicals. Further, they are effective deactivators of electronically excited sensitizer molecules which are involved in the generation of radicals and singlet oxygen [63, 64].

The interaction of carotenoids with $^1O_2$ depends largely on physical quenching which involves direct energy transfer between the molecules. The energy of singlet molecular oxygen is transferred to the carotenoid resulting in the formation of ground state oxygen and a triplet excited carotenoid. Dissipating its energy by interaction with the surrounding solvent, the carotenoid returns to its ground state and no further reactions take place. Because the carotenoids remain intact during physical quenching of $^1O_2$ or excited sensitizers, they can be reused severalfold in quenching cycles.

The efficacy of carotenoids for physical quenching is related to the number of conjugated double bonds present in the molecule which determines their lowest triplet energy level. Lutein, zeaxanthin, β-carotene and structurally related carotenoids have triplet energy levels close to that of $^1O_2$ thus, enabling efficient energy transfer. They are the most efficient naturally occurring quenchers for $^1O_2$ with quenching rate constants around $5–12 \times 10^9 M^{-1}s^{-1}$. However, the singlet oxygen quenching activity of carotenoids is dependent on the environment [65]. When incorporated into a model membrane, xanthophylls were less active than hydrocarbon carotenoids. In this system, lycopene and β-carotene exhibited the fastest singlet oxygen quenching rate constants whereas lutein and zeaxanthin were less efficient.

In contrast to physical quenching, chemical reactions between the excited oxygen and carotenoids are of minor importance, contributing less than 0.05%

to the total quenching rate. However, this process finally leads to the decomposition of the carotenoid, known as photobleaching. In vitro experiments have demonstrated that lutein and zeaxanthin are more stable than β-carotene and lycopene under photooxidative conditions. In a mixture of the four carotenoids the loss of lutein and zeaxanthin was less pronounced when the solution was irradiated with UV light in the presence of the sensitizer Rose Bengal [66]. The macular carotenoids were also more resistant towards irradiation with sunlight.

Among the various other ROS which are formed in the organism, carotenoids most efficiently scavenge peroxyl radicals. These are generated in the process of lipid peroxidation, and scavenging of this species interrupts the reaction sequence which would otherwise finally lead to damage of lipophilic compartments. Due to their lipophilicity and specific property to interact with peroxyl radicals, carotenoids are thought to play an important role in the protection of cellular membranes against oxidative damage [67]. The antioxidant activity of carotenoids regarding the deactivation of peroxyl radicals likely depends on the formation of radical adducts forming a resonance-stabilized carbon-centered radical.

A variety of products have been detected subsequent to oxidation of carotenoids, including carotenoid epoxides and apo-carotenoids of different chain length [68]. It should be noted that these compounds might possess biological activities and interfere with signaling pathways when present in unphysiologically high amounts [69].

The antioxidant activity of carotenoids depends on the oxygen tension present in the system [70, 71]. At low partial pressures of oxygen such as those found in most tissues under physiological conditions, β-carotene was found to inhibit oxidation. In contrast, prooxidant activities of carotenoids have been demonstrated in several in vitro experiments at high oxygen tension. However, it is still not known if prooxidant properties of carotenoids play a role in vivo.

A number of in vitro studies were carried out in order to compare the antioxidant activities of structurally different carotenoids. The results vary very much depending on the system used for investigation. The mechanism and rate of scavenging is strongly dependent on the nature of the oxidizing radical species and less dependent on the carotenoid structure [72, 73].

When the antioxidant activity of carotenoids was assayed in multilamellar liposomes measuring formation of thiobarbituric acid-reactive substances (TBARS) after challenge with 2,2′-azo-bis(2,4-dimethylvaleronitrile) (AMVN) the following ranking was determined: lycopene > α-carotene > β-cryptoxanthin > zeaxanthin = β-carotene > lutein [74]. In the TEAC assay, which investigates the potency to scavenge the ABTS radical, lutein and zeaxanthin exhibited comparable activities and were somewhat less efficient than β-carotene and lycopene [75].

In a model system using egg yolk lecithin liposomal membranes, UV-induced lipid oxidation was also slowed down by lutein and zeaxanthin. In this system, zeaxanthin appeared to be a better photoprotector during prolonged UV exposure. It was suggested that the differences in the protective efficacy of lutein and zeaxanthin in lipid membranes are attributable to a different organization of zeaxanthin-lipid and lutein-lipid membranes. Zeaxanthin was found to adopt roughly vertical orientation with respect to the plane of the membrane whereas the existence of two orthogonally oriented pools of lutein, one following the orientation of zeaxanthin and the second parallel with respect to the plane of the membrane was thought to play a role in photoprotection [76, 77].

There is evidence that the antioxidant effects of carotenoids depend on the concentration in the system with an optimal concentration for each compound. When human skin fibroblasts (loaded with single carotenoids) were exposed to UVB light, lycopene, β-carotene, and lutein were capable of decreasing UV-induced formation of TBARS, an indicator for lipid oxidation. The amounts of carotenoid needed for optimal protection were divergent: 0.05, 0.40, and 0.30 nmol/mg protein for lycopene, β-carotene, and lutein, respectively. At levels below the optimum, less protection was found whereas at higher levels prooxidant effects were observed [78].

Carotenoids are part of a complex antioxidant network, and it has been suggested that interactions between structurally different compounds with variable antioxidant activity provide additional protection against increased oxidative stress. For example, vitamin C, the most powerful water-soluble antioxidant in human blood and tissues, acts as a regenerator for vitamin E in lipid systems [79]. Synergistic interactions against UVA-induced photooxidative stress have been observed in cultured human fibroblasts when combinations of antioxidants were applied with β-carotene as the main component [80]. The antioxidant activity of carotenoid mixtures was assayed in multilamellar liposomes, measuring the inhibition of the formation of TBARS [74]. Mixtures were more effective than single compounds, and the synergistic effect was most pronounced when lycopene or lutein were present. The superior protection of mixtures may also be related to specific positioning of different carotenoids in membranes.

Carotenoids as Blue Light Filters

The photoreceptors in the retina are susceptible to damage by light, particularly blue light [81]. As already mentioned, pigments like lipofuscin may act as photosensitizers and have been considered to be involved in pathways leading to photooxidative damage. Upon irradiation with blue light, lipofuscin fluorophores mediate cellular damage and induce apoptosis [82].

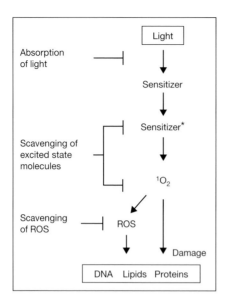

Fig. 2. Mechanisms of protection against photooxidative damage.

Based on the spectral properties of carotenoids, it has been postulated that one of the important tasks of lutein and zeaxanthin in the macula is filtering of blue light. However, all major dietary carotenoids including β-carotene and lycopene are efficient blue light filters in homogenous solution. Thus, it remains unclear why lutein and zeaxanthin should be preferably used as filtering compounds in the retina. It is known that spectral properties as well as antioxidant activities change with the environment. Therefore, the filtering effects of lutein and zeaxanthin in comparison to those of lycopene and β-carotene were investigated in membrane model using unilamellar liposomes [83]. Liposomes were loaded in the hydrophilic core space with a fluorescent dye, excitable by blue light, and various carotenoids were incorporated into the lipophilic membrane. The fluorescence emission in carotenoid-containing liposomes was lower than in controls when exposed to blue light, indicating a filter effect. At low concentrations, all carotenoids exhibited similar activities. However, the xanthophylls could be incorporated in higher amounts into the membrane and showed a better filtering efficacy than β-carotene or lycopene.

Conclusion

Evidence from epidemiology, animal studies, and in vitro experiments supports the hypothesis that the major macular pigments, lutein and zeaxanthin, protect the central retina against degenerative processes. The unique distribution, localization and the high levels of both carotenoids within the macula lutea add further evidence for a specific protective function in this tissue. Due to the physicochemical properties of carotenoids it is likely that the major task of the macula carotenoids is related to the protection of the central retina from photooxidative damage. Filtering blue light and scavenging ROS are most likely the mechanisms of protection (fig. 2). Upon supplementation, levels of lutein increase in the macula and high levels of intake are apparently related to a lower risk for AMD. One of the major tasks for the future will be to provide unequivocal evidence that an increased consumption of macular carotenoids helps to prevent AMD.

References

1   Olson JA, Krinsky NI: Introduction: The colorful fascinating world of the carotenoids. Important physiologic modulators. FASEB J 1995;9:1547–1550.
2   Demmig-Adams B, Adams WW III: Antioxidants in photosynthesis and human nutrition. Science 2002;298:2149–2153.
3   Pfander H: Key to Carotenoids. Basel, Birkhäuser, 1987.
4   Stahl W, Schwarz W, Sundquist AR, Sies H: cis-trans isomers of lycopene and beta-carotene in human serum and tissues. Arch Biochem Biophys 1992;294:173–177.
5   Khachik F, Spangler CJ, Smith JC, Canfield LM, Steck A, Pfander H: Identification, quantification, and relative concentrations of carotenoids and their metabolites in human milk and serum. Anal Chem 1997;69:1873–1881.
6   Van Duyn MA, Pivonka E: Overview of the health benefits of fruit and vegetable consumption for the dietetics professional: Selected literature. J Am Diet Assoc 2000;100:1511–1521.
7   Meydani M: Nutrition interventions in aging and age-associated disease. Ann NY Acad Sci 2001; 928:226–235.
8   Mayne ST: Beta-carotene, carotenoids, and disease prevention in humans. FASEB J 1996;10: 690–701.
9   Stahl W, Ale-Agha N, Polidori MC: Non-antioxidant properties of carotenoids. Biol Chem 2002; 383:553–558.
10 Stahl W, Sies H: Antioxidant activity of carotenoids. Mol Aspects Med 2003;24:345–351.
11 Krinsky NI, Landrum JT, Bone RA: Biologic mechanisms of the protective role of lutein and zeaxanthin in the eye. Annu Rev Nutr 2003;23:171–201.
12 Beatty S, Murray IJ, Henson DB, Carden D, Koh H, Boulton ME: Macular pigment and risk for age-related macular degeneration in subjects from a Northern European population. Invest Ophthalmol Vis Sci 2001;42:439–446.
13 Britton G, Liaaen-Jensen S, Pfander H: Carotenoids, Vol 1B: Spectroscopy. Basel, Birkhäuser, 1995.
14 Stahl W, Sundquist AR, Hanusch M, Schwarz W, Sies H: Separation of β-carotene and lycopene geometrical isomers in biological samples. Clin Chem 1993;39:810–814.
15 Sandmann G: Carotenoid biosynthesis and biotechnological application. Arch Biochem Biophys 2001;385:4–12.
16 Landrum JT, Bone RA: Lutein, zeaxanthin, and the macular pigment. Arch Biochem Biophys 2001;385:28–40.
17 Pelz R, Schmidt-Faber B, Heseker H: Carotenoid intake in the German National Food Consumption Survey (in German). Z Ernährungswiss 1998;37:319–327.
18 Johnson-Down L, Saudny-Unterberger H, Gray-Donald K: Food habits of Canadians: Lutein and lycopene intake in the Canadian population. J Am Diet Assoc 2002;102:988–991.

19 Breithaupt DE, Bamedi A: Carotenoids and carotenoid esters in potatoes (*Solanum tuberosum* L.): New insights into an ancient vegetable. J Agric Food Chem 2002;50:7175–7181.

20 Weller P, Breithaupt DE: Identification and quantification of zeaxanthin esters in plants using liquid chromatography-mass spectrometry. J Agric Food Chem 2003;51:7044–7049.

21 Breithaupt DE, Wirt U, Bamedi A: Differentiation between lutein monoester regioisomers and detection of lutein diesters from marigold flowers (*Tagetes erecta* L.) and several fruits by liquid chromatography-mass spectrometry. J Agric Food Chem 2002;50:66–70.

22 Brady WE, Mares-Perlman JA, Bowen P, Stacewicz-Sapuntzakis M: Human serum carotenoid concentrations are related to physiologic and life style factors. J Nutr 1996;126:129–137.

23 Olmedilla B, Granado F, Southon S, Wright AJ, Blanco I, Gil-Martinez E, et al: Serum concentrations of carotenoids and vitamins A, E, and C in control subjects from five European countries. Br J Nutr 2001;85:227–238.

24 Kaplan LA, Lau JM, Stein EA: Carotenoid composition, concentrations, and relationship in various human organs. Clin Physiol Biochem 1990;8:1–10.

25 Schmitz HH, Poor CL, Wellman RB, Erdman JW: Concentrations of selected carotenoids and vitamin A in human liver, kidney and lung tissue. J Nutr 1991;121:1613–1621.

26 Wingerath T, Sies H, Stahl W: Xanthophyll esters in human skin. Arch Biochem Biophys 1998; 355:271–274.

27 Roodenburg AJC, Leenen R, van het Hof KH, Weststrate JA, Tijburg LBM: Amount of fat in the diet affects bioavailability of lutein esters but not of α-carotene, β-carotene, and vitamin E in humans. Am J Clin Nutr 2000;71:1187–1193.

28 Riedl J, Linseisen J, Hoffmann J, Wolfram G: Some dietary fibers reduce the absorption of carotenoids in women. J Nutr 1999;129:2170–2176.

29 Weststrate JA, van het Hof KH: Sucrose polyester and plasma carotenoid concentrations in healthy subjects. Am J Clin Nutr 1995;62:591–597.

30 Van het Hof KH, Gärtner C, West CE, Tijburg BM: Potential of vegetable processing to increase the delivery of carotenoids to man. Int J Vitam Nutr Res 1998;68:366–370.

31 Van het Hof KH, Brouwer IA, West CE, Haddeman E, Steegers-Theunissen RPM, van Dusseldorp M, et al: Bioavailability of lutein from vegetables is five times higher than that of β-carotene. Am J Clin Nutr 1999;70:261–268.

32 Gärtner C, Stahl W, Sies H: Preferential increase in chylomicron levels of the xanthophylls lutein and zeaxanthin compared to β-carotene in the human. Int J Vitam Nutr Res 1996;66:119–125.

33 Van den Berg H: Carotenoid interactions. Nutr Rev 1999;57:1–10.

34 Van den Berg H, van Vliet T: Effect of simultaneous, single oral doses of β-carotene with lutein or lycopene on the β-carotene and retinyl ester responses in the triacylglycerol-rich lipoprotein fraction of men. Am J Clin Nutr 1998;68:82–89.

35 Bowen PE, Herbst-Espinosa SM, Hussain EA, Stacewicz-Sapuntzakis M: Esterification does not impair lutein bioavailability in humans. J Nutr 2002;132:3668–3673.

36 Breithaupt DE, Weller P, Wolters M, Hahn A: Comparison of plasma responses in human subjects after the ingestion of 3R,3R′-zeaxanthin dipalmitate from wolfberry (*Lycium barbarum*) and non-esterified 3R,3R′-zeaxanthin using chiral high-performance liquid chromatography. Br J Nutr 2004;91:707–713.

37 Breithaupt DE, Weller P, Wolters M, Hahn A: Plasma response to a single dose of dietary beta-cryptoxanthin esters from papaya (*Carica papaya* L.) or non-esterified β-cryptoxanthin in adult human subjects: A comparative study. Br J Nutr 2003;90:795–801.

38 Granado F, Olmedilla B, Blanco I, Rojas-Hidalgo E: Major fruit and vegetable contributors to the main serum carotenoids in the Spanish diet. Eur J Clin Nutr 1996;50:246–250.

39 Bernstein PS, Khachik F, Carvalho LS, Muir GJ, Zhao DY, Katz NB: Identification and quantitation of carotenoids and their metabolites in the tissues of the human eye. Exp Eye Res 2001;72: 215–223.

40 Bone RA, Landrum JT, Tarsis SL: Preliminary identification of the human macular pigment. Vision Res 1985;11:1531–1535.

41 Handelman GJ, Dratz EA, Reay CC, van Kuijk FJGM: Carotenoids in the human macula and whole retina. Invest Ophthalmol Vis Sci 1988;29:850–855.

42 Khachik F, Bernstein PS, Garland DL: Identification of lutein and zeaxanthin oxidation products in human and monkey retinas. Invest Ophthalmol Vis Sci 1997;38:1802–1811.

43 Schalch W, Dayhaw-Barker P, Barker FM: The carotenoids of the human retina; in Taylor A (ed): Nutritional and Environmental Influences on the Eye. Boca Raton, CRC Press, 1999, pp 215–250.

44 Landrum JT, Bone RA, Chen Y, Herrero C, Llerena CM, Twarowska E: Carotenoids in the human retina. Pure Appl Chem 1999;71:2237–2244.

45 Bone RA, Landrum JT, Guerra LH, Ruiz CA: Lutein and zeaxanthin dietary supplements raise macular pigment density and serum concentrations of these carotenoids in humans. J Nutr 2003; 133:992–998.

46 Bone RA, Landrum JT, Hime GW, Cains A, Zamor J: Stereochemistry of the human macular carotenoids. Invest Ophthalmol Vis Sci 1993;34:2033–2040.

47 Khachik F, de Moura FF, Zhao DY, Aebischer CP, Bernstein PS: Transformations of selected carotenoids in plasma, liver, and ocular tissues of humans and in nonprimate animal models. Invest Ophthalmol Vis Sci 2002;43:3383–3392.

48 Krinsky NI: Possible biologic mechanisms for a protective role of xanthophylls. J Nutr 2002;132: 540S–542S.

49 Bernstein PS, Balashov NA, Tsong ED, Rando RR: Retinal tubulin binds macular carotenoids. Invest Ophthalmol Vis Sci 1997;38:167–175.

50 Sommerburg O, Siems WG, Hurst JS, Lewis JW, Kliger DS, van Kuijk FJGM: Lutein and zeaxanthin are associated with photoreceptors in the human retina. Curr Eye Res 1999;19:491–495.

51 Sies H: Biochemistry of oxidative stress. Angew Chem Int Ed Engl 1986;25:1058–1071.

52 Sies H: Oxidative stress: From basic research to clinical application. Am J Med 1991;91:31S–38S.

53 Halliwell B, Gutteridge JMC: Free Radicals in Biology and Medicine, ed 3. Oxford, Clarendon Press, 1999.

54 Sies H, Stahl W: Antioxidants and human health; in Paoletti P, Sies H, Bug J, Grossi E, Poli A (eds): Vitamin C: The State of the Art in Disease Prevention Sixty Years after Nobel Prize. Berlin, Springer, 1998, pp 1–11.

55 Sies H: Antioxidants in Disease Mechanisms and Therapy. London, Academic Press, 1997.

56 Sies H: What is oxidative stress? in Keaney JF (ed): Oxidative Stress and Vascular Disease. Boston, Kluwer Academic, 2000, pp 1–8.

57 Davies MJ, Truscott RJW: Photo-oxidation of proteins and its consequences; in Giacomoni PU (ed): Sun Protection in Man. Amsterdam, Elsevier, 2001, pp 251–275.

58 Girotti AW: Lipid photooxidative damage in biological membranes: Reaction mechanisms, cytotoxic consequences, and defense strategies; in Giacomoni PU (ed): Sun Protection in Man. Amsterdam, Elsevier, 2001, pp 231–250.

59 Ravanat JL, Douki T, Cadet J: UV damage to nucleic acid components; in Giacomoni PU (ed): Sun Protection in Man. Amsterdam, Elsevier, 2001, pp 207–230.

60 Davies MJ: Singlet oxygen-mediated damage to proteins and its consequences. Biochem Biophys Res Commun 2003;305:761–770.

61 Lamb LE, Simon JD: A2E: A component of ocular lipofuscin. Photochem Photobiol 2004;79: 127–136.

62 Wassell J, Davies S, Bardsley W, Boulton M. The photoreactivity of the retinal age pigment lipofuscin. J Biol Chem 1999;274:23828–23832.

63 Truscott TG: The photophysics and photochemistry of the carotenoids. J Photochem Photobiol B Biol 1990;6:359–371.

64 Young AJ, Lowe GM: Antioxidant and prooxidant properties of carotenoids. Arch Biochem Biophys 2001;385:20–27.

65 Cantrell A, McGarvey DJ, Truscott TG, Rancan F, Bohm F: Singlet oxygen quenching by dietary carotenoids in a model membrane environment. Arch Biochem Biophys 2003;412:47–54.

66 Siems WG, Sommerburg O, van Kuijk FJ: Lycopene and β-carotene decompose more rapidly than lutein and zeaxanthin upon exposure to various pro-oxidants in vitro. Biofactors 1999;10:105–113.

67 Sies H, Stahl W: Vitamins E and C, β-carotene, and other carotenoids as antioxidants. Am J Clin Nutr 1995;62:1315S–1321S.

68 Kennedy TA, Liebler DC: Peroxyl radical oxidation of β-carotene: Formation of β-carotene epoxides. Chem Res Toxicol 1991;4:290–295.

69 Wang XD, Russell RM: Procarcinogenic and anticarcinogenic effects of β-carotene. Nutr Rev 1999;57:263–272.

70   Burton GW, Ingold KU: β-Carotene: An unusual type of lipid antioxidant. Science 1984;224: 569–573.

71   Palozza P: Prooxidant actions of carotenoids in biologic systems. Nutr Rev 1998;56:257–265.

72   Mortensen A, Skibsted LH, Sampson J, Rice-Evans CA, Everett SA: Comparative mechanisms and rates of free radical scavenging by carotenoid antioxidants. FEBS Lett 1997;418:91–97.

73   Mortensen A, Skibsted LH, Truscott TG: The interaction of dietary carotenoids with radical species. Arch Biochem Biophys 2001;385:13–19.

74   Stahl W, Junghans A, de Boer B, Driomina E, Briviba K, Sies H: Carotenoid mixtures protect multilamellar liposomes against oxidative damage: Synergistic effects of lycopene and lutein. FEBS Lett 1998;427:305–308.

75   Miller NJ, Sampson J, Candeias LP, Bramley PM, Rice-Evans CA: Antioxidant activities of carotenes and xanthophylls. FEBS Lett 1996;384:240–242.

76   Gruszecki WI: Carotenoids in membranes; in Frank HA, Young AJ, Britton G, Cogdell RJ (eds): The Photochemistry of Carotenoids. Dordrecht, Kluwer Academics, 1999, pp 363–379.

77   Sujak A, Gabrielska J, Grudzinski W, Borc R, Mazurek P, Gruszecki WI: Lutein and zeaxanthin as protectors of lipid membranes against oxidative damage: The structural aspects. Arch Biochem Biophys 1999;371:301–307.

78   Eichler O, Sies H, Stahl W: Divergent optimum levels of lycopene, β-carotene and lutein protecting against UVB irradiation in human fibroblasts. Photochem Photobiol 2002;75:503–506.

79   Niki E, Noguchi N, Tsuchihashi H, Gotoh N: Interaction among vitamin C, vitamin E, and β-carotene. Am J Clin Nutr 1995;62:1322S–1326S.

80   Böhm F, Edge R, Lange L, Truscott TG: Enhanced protection of human cells against ultraviolet light by antioxidant combinations involving dietary carotenoids. J Photochem Photobiol B Biol 1998;44:211–215.

81   Ham WT, Mueller HA: The photopathology and nature of the blue light and near-UV retinal lesions produced by lasers and other optical sources; in Wolbarsht ML (ed): Laser Applications in Medicine and Biology. New York, Plenum Press, 1989, pp 191–246.

82   Shaban H, Richter C: A2E and blue light in the retina: The paradigm of age-related macular degeneration. Biol Chem 2002;383:537–545.

83   Junghans A, Sies H, Stahl W: Macular pigments lutein and zeaxanthin as blue light filters studied in liposomes. Arch Biochem Biophys 2001;391:160–164.

84   Olmedilla B, Granado F, Blanco I, Rojas-Hidalgo E: Seasonal and sex-related variations in six serum carotenoids, retinol, and α-tocopherol. Am J Clin Nutr 1994;60:106–110.

85   Cantilena LR, Nierenberg DW: Simultaneous analysis of five carotenoids in human plasma by isocratic high-performance liquid chromatography. J Micronutr Anal 1989;6:127–145.

86   Sowell AL, Huff DL, Yeager PR, Caudill SP, Gunter EW: Retinol, α-tocopherol, lutein/zeaxanthin, β-cryptoxanthin, lycopene, α-carotene, trans-β-carotene, and four retinyl esters in serum determined simultaneously by reversed-phase HPLC with multiwavelength detection. Clin Chem 1994;40:411–416.

87   Forman MR, Lanza E, Yong LC, Holden JM, Graubard BI, Beecher GR, et al: The correlation between two dietary assessments of carotenoid intake and plasma carotenoid concentrations: Application of a carotenoid food-composition database. Am J Clin Nutr 1993;58:519–524.

88   Yong LC, Forman MR, Beecher GR, Graubard BI, Campbell WS, Reichmann ME, et al: Relationship between dietary intake and plasma concentrations of carotenoids in premenopausal women: Application of the USDA-NCI carotenoid food-composition database. Am J Clin Nutr 1994;60:223–230.

Prof. Dr. W. Stahl
Heinrich Heine University Düsseldorf
Institute of Biochemistry and Molecular Biology I
PO Box 101007, DE–40001 Düsseldorf (Germany)
Tel. +49 211 8112711, Fax +49 211 8113029, E-Mail wilhelm.stahl@uni-duesseldorf.de

Augustin A (ed): Nutrition and the Eye.
Dev Ophthalmol. Basel, Karger, 2005, vol 38, pp 89–102

··························

Selenium, Selenoproteins and Vision

Leopold Flohé

MOLISA GmbH, Magdeburg, Germany

Abstract

Selenium biochemistry is reviewed in respect to its presumed relevance to age-related ocular diseases. Selenium is an essential trace element that exerts its physiological role as selenocysteine residue in at least 25 distinct selenoenzymes in mammals. Lack of GPx-1 due to alimentary selenium deprivation has been inferred to induce cataract in rats and was demonstrated to cause cataracts in mice by targeted gene disruption. The role of other selenoproteins in the eye remains to be worked out. Selenium in excess of the tiny amounts required for selenoprotein synthesis is toxic in general and causes cataracts in experimental animals. Clinical evidence for a protective role of selenium in the development of cataract, macula degeneration, retinitis pigmentosa or any other ocular disease is not available, likely because suboptimum selenium intake, as it may result from unbalanced diet, does not cause any pathologically relevant selenium deficiency in the eye. At present, there is neither theoretical nor an empirical basis to expect beneficial effects of selenium supplementation beyond the dietary reference intakes of 55 µg/day in the context of ocular diseases.

In the context of ocular diseases, selenium is most frequently quoted as an agent that causes cataracts in experimental rodents. However, deficient alimentary supply of the essential trace element is also implicated in the development of cataracts [1, 2]. The latter effect is commonly discussed to be related to an antioxidant action of selenium. Since oxidative damage is also believed to contribute to age-related macula degeneration and retinitis pigmentosa, the presumed antioxidant selenium might equally be relevant to these diseases. Supportive experimental or clinical data, however, are scarce, and the seemingly conflicting findings demand a critical re-evaluation that is based on solid knowledge of the biological roles of selenium.

This article will therefore briefly summarize relevant aspects of selenium biochemistry in mammals, compile the knowledge on selenoproteins with special emphasis on those present in the eye, try to explain experimental or clinical data by established molecular events and finally line out what should reasonably be considered to become clinically important.

Unspecific Selenium Effects versus Enzymatic Selenium Catalysis

Selenium exerts its beneficial biological role as constituent of an estimated total of 25 distinct proteins (table 1) [3]. They comprise five thiol-dependent peroxidases, commonly called glutathione peroxidases (GPx), three deiodinases (DI), which are involved in the synthesis and degradation of the thyroid hormones, three thioredoxin reductases (TR), the selenium transport protein (SelPP), the selenophosphate synthetase that is required for the synthesis of all other selenoproteins, and a variety of further proteins known by deduced amino acid sequence, the biological role of which is still poorly defined [4].

In these proteins, selenium is present as one or more selenocysteine residues that are integrated into the amino acid chains at specific positions. The specific incorporation of selenocysteine into the proteins is determined by a complex coding mechanism, wherein the stop codon TGA is recoded by means of a secondary mRNA structure called SECIS (for selenocysteine incorporation sequence) and the pertinent translation factors, SBP-2 and mSelB. The former recognizes the SECIS, the latter a specific selenocysteyl-loaded tRNA$^{(ser)sec}$. Interestingly, charging of tRNA$^{(ser)sec}$ differs from the common pathway; the tRNA has first to be loaded with serine. The seryl residue is then transformed into a selenocysteyl residue by selenocysteine synthase with selenophosphate as substrate. If the charged selenocysteyl-tRNA$^{(ser)sec}$ is not sufficiently available, the stop codon nature of TGA becomes dominant again despite the presence of the SECIS in the particular mRNA, that means the 'selenoprotein' is truncated at the position where selenocysteine was to be inserted [reviewed in 4].

An important phenomenon to understand the biological consequences of selenium shortage is the 'hierarchy of selenoproteins'. The term describes the observation that the individual selenoproteins respond differently to selenium restriction. Out of the well-investigated selenoproteins, the classical glutathione peroxidase, GPx-1, and GPx-3, the extracellular variant, decline most readily in selenium deficiency and recover slowly upon re-supplementation. In contrast, GPx-2, the gastrointestinal form, and phospholipid hydroperoxide GPx (GPx-4) remain reasonably high even in moderate to severe selenium deficiency, the remaining selenoproteins ranking in between. The underlying molecular mechanism is not completely understood. One of the reasons of the fast decline and slow recovery of GPx-1 and GPx-3 is a degradation of the pertinent mRNAs in

Table 1. Selenoproteins 2003 [3]

Mammalian selenoproteins	Common abbreviations
Glutathione peroxidase	GPx
Cytosolic or classical GPx	cGPx, GPx-1
Phospholipid hydroperoxide GPx	PHGPx, GPx-4
Plasma GPx	pGPx, GPx-3
Gastrointestinal GPx	GI-GPx, GPx-2
GPx3-homolog	GPx-6
Iodothyronine deiodinases	
5′-deiodinase, type 1	5′DI-1
5′-deiodinase, type 2	5′DI-2
5-deiodinase, type 3	5-DI-3
Thioredoxin reductases	TR
Thioredoxin reductase	TR-2
Mitochondrial thioredoxin reductase	SelZf1
Thioredoxin reductase homologs	SelZf2
Selenophosphate synthetase-2	SPS2
15-kDa selenoprotein (T cells)	
Selenoprotein P	SelP
Selenoprotein W	SelW
Selenoprotein R (methionine sulfoxide reductase)	MrsB
Selenoprotein T	SelT
Selenoprotein M	SelM
Selenoprotein N (knockout causes muscular dystrophy with spinal rigidity and restrictive respiratory syndrome)	SelN
Selenoprotein H	
Selenoprotein I	
Selenoprotein K	
Selenoprotein O	
Selenoprotein S	
Selenoprotein V	

response to selenium deprivation. A selenium-dependent affinity shift of RNA-binding proteins likely contributes to the differential mRNA stabilities. Out of the proteins of the selenoprotein machinery only SBP-2 might function as the required selenium sensor, as it only binds tRNA$^{(ser)sec}$ if this is charged with selenocysteine. However, since SBP-2 does not directly interact with the mRNA, it would have to cooperate with an RNA-binding protein such as mSelB

to signal the selenium status to the mRNA/protein complex, which is also called the selenosome [reviewed in 4, 5].

Equally important is the differential delivery of selenium to particular tissues. Privileged organs that retain their selenium status in pronounced selenium deficiency are thyroid, brain and testis. The molecular basis of this up to recently mysterious phenomenon is now emerging. Food-derived selenium is incorporated in the liver into SelPP, a protein with up to 17 selenocysteine residues (10 in humans). SelPP is secreted into the circulation apparently to deliver its selenium to sites of particular demand. The present hypothesis is that SelPP binds to a receptor in the privileged tissues, is internalized there and degraded to provide selenium for de novo synthesis of selenoproteins. This view is corroborated by a dramatic drop of selenium content in privileged tissues associated with elevated selenium levels in livers of SelPP knockout mice [6]. Unfortunately, it has so far not been investigated whether ocular tissues are similarly supplied with selenium, as is the brain.

It remains to be discussed what happens to adsorbed selenium beyond the amount required for the synthesis of selenoproteins. To some extent this depends on the chemical nature of the particular seleno compound. Selenomethionine, for instance, is stochastically incorporated into proteins instead of methionine, the consequences thereof being unclear. As a rule, however, bioavailable selenium, be it selenite from drinking water or selenoamino acids from meat or fish protein, are transformed to the same intermediate, selenide. In case of selenite, the reduction can be achieved directly by any of the thioredoxin reductases or by reaction with GSH to selenaglutathione trisulfide and reduction thereof by glutathione reductase or thioredoxin reductases. The selenoamino acids are transformed by the transulfuration pathway and, ultimately, H_2Se is released from selenocysteine by (seleno)cysteine lyase [4]. H_2Se serves as the precursor of selenophosphate, which is used for selenoprotein synthesis, as outlined above. Any excess beyond the tiny genetically determined demand needs to be disposed immediately, because seleno compounds are highly reactive and accordingly toxic and no storage mechanism, apart from the limited capacity of SelPP synthesis, is known in mammals. The first line of defense against excess selenium is SAM-dependent methylation of H_2Se. It yields the volatile mono- and dimethylselenium derivatives that are exhaled and account for the smell of rotten horseradish or garlic that is typical of acute selenium poisoning. These volatile metabolites easily penetrate the blood-brain barrier and are the main culprits of selenium's neurotoxicity. Further methylation yields the trimethyl-selenonium ion that is excreted with the urine. Once the methylation capacity is exhausted, which usually results from chronic overexposure to selenium, the element starts to disclose it so-called antioxidant potential – with disastrous consequences [more details reviewed in 2, 4, 7].

For sake of clarity it has to be stressed that selenium is not an antioxidant, either in the chemical meaning of this term or in a biological sense. As will be discussed below, selenium plays an important role in the metabolism of H_2O_2 and other hydroperoxides by being a functional heteroatom in peroxidases and thereby contributes to the prevention of free radical formation and related tissues damage. But this does not justify its mislabeling as an 'antioxidant'. In chemical terminology, an antioxidant is a compound that reacts with free radicals, thereby becomes transformed into a less reactive radical itself, and thus slows down or terminates free radical chain reactions. None of the selenium compounds contained in food or metabolites thereof meets these characteristics. More importantly, the mass law implies that the efficacy of an antioxidant increases with its concentration. What instead happens in 'supranutritional dosing' of selenium is easily predicted and has been amply verified experimentally [8]: The excess selenium ends up in a pool of selenium compounds of the oxidation state -2, i.e. selenides or selenols. Being strong reductants, such compounds react with the most abundant oxidant, i.e. molecular dioxygen (O_2), with formation of superoxide anion radicals and hydrogen peroxide (H_2O_2). The seleno compounds thereby oxidized are enzymatically reduced as outlined above, and undergo autoxidation again. In short, any selenium surpassing the biological needs and the very limited storage capacity starts redox cycling, which is the most efficient way to cause oxidative damage in biological systems.

These basic principles of selenium biochemistry disclose why the therapeutic window of any selenium compound is inevitably small. They further reveal why the symptoms of chronic selenium intoxication, which is associated with the oxidative stress markers typical of redox cyclers, often resemble those of selenium deficiency, which results in impaired detoxification of hydroperoxides.

Selenoproteins of the Eye

The majority of the mammalian selenoproteins were discovered in the past decade. Accordingly, understanding of their biology is mostly limited, and their presence in ocular tissues has been verified in exceptional cases only.

It must be inferred that the eye contains cytosolic (TR1) and mitochondrial (TR3) thioredoxin reductases that are indispensable for ribonucleotide reduction via thioredoxin and also determine other functions of the pleiotropic redox mediators of the thioredoxin family. Similarly, each cell has to be equipped with selenophosphate synthetase to synthesize pivotal selenoproteins such as the thioredoxin reductases. Selenoprotein W prevails in muscle and nervous tissue but its presence in the eye remains to be established. Virtually nothing is

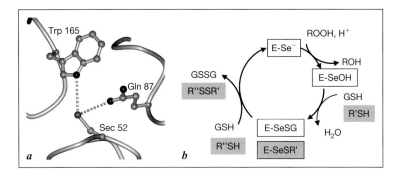

Fig. 1. The glutathione peroxidase reaction. *a* The catalytic triad that is strictly conserved by the whole GPx superfamily with the only exception that selenocysteine may be replaced by cysteine, which is associated with low peroxidases activity. Residue numbers of bovine GPx-1 are given as example. *b* The catalytic cycle of GPx that is essentially characterized by redox shuttling of its selenium moiety. In case of GPx-1, GSH serves as reducing substrate, other isoenzymes accept different thiols (RSH).

known about iodothyronine deiodinases (DI) in the eye. They regulate local thyroid hormone activity by generating the active 3,5,3′-triiodothyronine from thyroxin (DI-1 and DI-2) or degrading thyroxin and 3,5,3′-triiodothyronine to reverse T3 (3,3′,5′-triiodothyronine) or other inactive compounds, respectively (DI-1 and DI-3), and their potential relevance to ocular affections in thyroid disturbances such as Graves' disease would certainly be of interest.

The only two selenoproteins that have unambiguously been shown to be present in the eye are GPx-1 and GPx-3. As early as 1965, i.e. long before the selenoprotein nature of any mammalian enzyme was recognized [9], the 'classic' cytosolic glutathione peroxidase (GPx-1) was isolated from lens by the pioneer of eye biochemistry, Antoinette Pirie [10]. The extracellular form GPx-3, which is primarily derived from the kidney, was found to be also synthesized in the ciliary body [11] and to be released into the aqueous humor [12]. Like all other members of the GPx family, these two enzymes reduce H_2O_2, organic hydroperoxides and peroxynitrite at the expense of thiols with high efficiency. Depending on the nature of the hydroperoxide, the bimolecular rate constants for the reaction of reduced enzyme with ROOH ranges between some 10^6 and 10^8 $M^{-1}s^{-1}$. These extreme rate constants depend on the selenium moiety, which forms a catalytic triad consisting of a selenocysteine, a glutamine and a tryptophan residue wherein the selenol function of the selenocysteine is dissociated and polarized for nucleophilic attack on peroxo groups [4] (fig. 1). In case of GPx-1, the reducing substrate is glutathione (GSH), whereas GPx-3 also accepts thioredoxin and glutaredoxin as reductants. The metabolic context of GPx-1 is

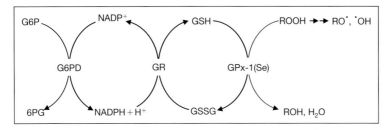

Fig. 2. Main metabolic context of GPx-1. The selenoprotein is typically supplied with reduction equivalents (NADPH) from the pentose phosphate shunt. Deficiencies in this pathway may impair regeneration of GSH and, in consequence, hydroperoxide (ROOH) detoxification via GPx-1. Accumulating hydroperoxide, by decomposing into alkoxyl (RO·) and hydroxyl radical (·OH), may then initiate free radical chain reactions.

straightforward. The reducing substrate GSH is regenerated by glutathione reductase, which predominantly receives its reduction equivalents as NADPH from the pentose-phosphate shunt (fig. 2). Instead, GPx-3 depends on the tiny amounts of thiols that are released into the extracellular space and has therefore been addressed as 'orphan enzyme' [7]. The lack of any known thiol-regenerating system in the extracellular space limits the capacity of GPx-3 to cope with an extensive hydroperoxide challenge. The different localization of these otherwise similar enzymes thus points to distinct biological roles: While GPx-1 has been established as the most important device of hydroperoxide detoxification in general, GPx-3 may regulate the extracellular peroxide tone that is implicated in the biosynthesis of inflammatory lipid mediators by lipoxygenases and may affect other signaling cascades [reviewed in 5, 7].

Glutathione Peroxidases and Experimental Cataracts

The interest of ophthalmologists in the relationship of the glutathione system and cataract development dates back to the 1950s of the last century and was extensively reviewed by Kinoshita [13] already in 1964. The observations were: (i) the lens has been reported to have a GSH content that, with about 10 mM, surpasses that of any other tissue [14]; (ii) GSH of the lens drops with age [15]; (iii) it is even more decreased in cataractous lenses [15], where (iv) glutathionylated proteins increase [15].

The first to recognize the link of these findings to H_2O_2 detoxification was evidently Antoinette Pirie, who not only identified GPx-1 in the lens but simultaneously presented a source of H_2O_2 that attacks the lens from the aqueous humor, where it is formed by autoxidation of another 'antioxidant', ascorbate [10].

The role of H_2O_2 in inducing cataract was then corroborated by Srivastava and Beutler [16] by incubating rabbit lenses with tyrosine and tyrosinase, which produces superoxide and/or H_2O_2 as by-product. Preceding cataract development GSH became oxidized and released to the medium as GSSG. Interestingly, a similar loss of GSH in the lens and export of GSSG to the aqueous humor was observed upon naphthalene feeding to rabbits [17], which likely is mediated by a metabolite of naphthalene, 1,2-naphthoquinone. The latter is a known redox cycler which oxidizes GSH by continuously generating H_2O_2 but aggravates the loss of GSH by reacting to covalent adducts [18]. Analogous reactions of 1,2-naphthoquinone with SH groups of β- and γ-crystallin, that are favored at low GSH concentrations, lead to insoluble colored proteins that contribute to cataract manifestations.

The latter findings reveal that H_2O_2 itself is not necessarily the agent that induces the cataractogenic protein modification. H_2O_2-derived free radicals may be the main culprits, and oxygen-centered radicals may be directly formed in the eye, e.g. by UV irradiation. To mimic such radical damage, rats were poisoned with the herbicide diquate. This compound is reduced univalently by NADPH and induces cataract via a radical that does not cause any significant loss of GSH [19]. Further evidence for GSH-independent cataracts is provided by genetics. A dominant cataract mutant (Nop/+) did not display any abnormalities in the enzymes related to the glutathione redox balance. In particular the activities of GPx, glutathione reductase and glucose-6-phosphate dehydrogenase were not affected [20]. Similarly, mutations in the Huntingtin interacting protein were found to be associated with cataracts [21].

On the other hand, inverse genetics finally corroborate the early hypotheses on hydroperoxides being key players in cataract development and the GSH system being protective: GPx-1$^{-/-}$ mice spontaneously develop complete lens opacification at an age of 15 months. This is preceded by progressive nuclear light scattering, distortion of fiber membranes and distension of interfiber space starting at 3 weeks of age and lamellar cataracts between 6 and 10 months [22]. This finding is of particular interest, since it represents the so far only phenotype observed in unstressed GPx-1$^{-/-}$ mice. These animals grow normally if not exposed to hydroperoxides, bacterial toxins that trigger an oxidative burst in phagocytes, redox-cycling herbicides or viral infections [reviewed in 7]. It has therefore to be concluded that the lens, in contrast to other tissues, is physiologically exposed to a certain oxidative stress that needs continuous protection by the selenoprotein GPx-1. In short, the GSH-dependent hydroperoxide detoxification may thus be rated as one of the genetically validated systems that protect against age-related cataract formation. The knockout experiments also provide a rational to explain the early observations of cataract development in selenium-deprived rodents [23, 24], which thus is likely the consequence of GPx-1 deficiency.

How then does the proven protection by selenium against cataract formation comply with the cataractogenic potential of the very same element? Cataracts can consistently be induced in young rats, rabbits and guinea pigs by selenite at dosages below the threshold causing acute systemic toxicity. The narrow time window of susceptibility and the comparatively low dosages suggested specific effects of selenite on the lens, and indeed a variety of phenomena have been observed in this cataract model that are not easily explained by unspecific selenium toxicity, e.g., a dramatic increase in calcium and phosphate in the lens, binding of radioactive selenium to lens proteins [compiled in 2], impairment of protein tyrosine phosphorylation and phosphatidylinositol-3-kinase activity [25], and calcium-induced proteolysis of β-crystallin [26]. However, whatever the seemingly specific mechanism of lens toxicity may be, all these effects are observed at dosages that surpass the ones required for optimum production of selenoproteins by more than three orders of magnitude: several milligrams/kilogram instead of 1 μg/kg. Under these conditions, selenium has clearly changed its face from an essential micronutrient to a pro-oxidative redox cycler. Accordingly, the total reducing capacity of the lens is decreased, as is evident from a decrease in GSH, NADPH and total protein sulfhydryls, and markers of oxidative damage such as malondialdehyde are substantially increased [2]. It therefore may be doubted if the seemingly specific effects of selenium reflect more than sites of the young lenses that are particularly prone to oxidative damage. In line with this view, selenite-induced cataract could be inhibited in vitro and in vivo with antioxidant extracts of green tea (*Camellia sinensis*) [27] or other antioxidants [28]. In conclusion, as an essential trace element up to daily dosages of 1 μg/kg body weight, selenium prevents cataract formation by guaranteeing optimum H_2O_2 detoxification via GPx-1. If given in marked excess, it induces cataracts by causing oxidative damage to the lens.

Animal Experiments versus Clinical Experience

The unequivocal demonstration of the indispensability of GPx-1 in the eye of mice justifies discussing the potential impact of the selenoenzyme in respect to ophthalmic diseases believed to be caused or aggravated by oxidative stress. Since GPx-1 belongs to the group of selenoproteins that declines rapidly in selenium deficiency, the problem addressed is intimately related to the question if selenium supplementation can ameliorate such diseases. To state it right away, supportive clinical evidence is surprisingly scarce.

To our knowledge there is not a single report that relates a genetic deficiency of GPx-1 or of any other selenoprotein to ocular diseases. Congenital cataract was reported in 1 out of 14 cases of heredited deficiency of GSH biosynthesis [27]. The evidence for a role of GPx-1 in the human eye becomes slightly more persuasive if the GSH-regenerating system GPx-1 depends on is considered. Glucose-6-phosphate dehydrogenase (G-6-PD) deficiency has for long been implicated as a risk factor for cataract formation [29], but the attempts to statistically verify an association of G-6-PD deficiency and cataract incidence remained largely disappointing [30, 31]. The results varied between geographical regions and, likely, between different types of G-6-PD mutations. Nevertheless, a trend towards higher cataract incidence was observed at least in younger or presenile patients [30, 32–34]. Likely the genetic deficiencies in GSH regeneration remain silent as long as the subjects are not challenged by pro-oxidant agents, as they do in respect to associated hematological disorders. A statistically verified association of cataract incidence and selenium deficiency is equally missing and has not even been reported for countries where selenium deficiency syndromes such as Keshan disease or Kashin-Beck disease were endemic. At first glance, this surprises in view of the experimental background. It may be revealing, however, that cataract induction in rats required second-generation selenium deprivation. This implies that either the selenium deficiency must be extreme and has to be sustained for years to mimic the GPx-1 status of GPx-1$^{-/-}$ mice or that the lenses of rats and man, respectively, are less prone to oxidative damage. Evidently a critically low selenium status of the eye is hardly achieved even at low, and not at all at suboptimum supply, because the eye is likely as privileged in selenium supply as the nervous tissue in general.

The second ophthalmic disease believed to result from, or to be aggravated by, oxidative damage, is retinitis pigmentosa (RP). The pigmented deposits reminding of lipofuscin are considered to result from co-oxidation of unsaturated lipid and proteins. The most convincing clinical support of the oxidative damage hypothesis of RP is the almost consistent association of the disease with untreated genetic deficiency of the α-tocopherol transfer protein that leads to a dramatic decrease of vitamin E in nervous tissue, ataxia and mental retardation [35]. The preventive efficacy of megadoses of vitamin E (up to 2 g/day) is commonly attributed to its antioxidant capacity that essentially consists in scavenging peroxyl radicals of lipids. The products of this reaction, hydroperoxides of fatty acids or complex lipids are substrates of GPx's, in particular of GPx-4. This biochemical link is often quoted to explain the synergism of selenium and vitamin E in preventing experimental deficiency syndromes [2, 7]. However, GPx-4, which is the isoenzyme specialized for the prevention of lipid peroxidation, ranks high in the hierarchy of selenoproteins, will not markedly decrease, unless selenium deficiency is extreme. The few clinical studies addressing this problem did not suggest any link between RP and selenium deficiency. Unexpectedly, GPx-1, which readily declines in selenium deficiency, was found to be elevated in RP [36, 37], a finding that, in principle, corroborates the oxidative stress hypothesis for RP, but rules out the contribution of selenium deficiency as a common etiological factor.

For age-related macular degeneration (ARMD) finally, the assumption of a radiation-triggered oxidative damage is theoretically most appealing. The invoked oxidative processes, however, are initiated by singlet oxygen, which is more readily quenched by the carotenoids of the *macula lutea* than by any other non-enzymatic or enzymatic process. Accordingly, the *macula lutea* protects itself against potential damage by the focused light by accumulation of carotenoids, specifically lutein and zeaxanthin, which are responsible for its typical color [38]. The protective role of other antioxidants or enzymatic mechanism is less clear. A study on 18 ARMD patients showed a lower blood glutathione reductase activity but comparable GPx activity when compared to age-matched controls [39]. More recently, a genome scan for ARMD-related markers identified a region on chromosome 5, where GPx-3 is located [40]. Altogether, however, the evidence for an impact of the selenium status to ARMD is weak.

Unfortunately, clinical trials with patient numbers that promise statistical power have preferentially been performed with complex mixtures of minerals and vitamins that are believed to act primarily as antioxidants. Despite the high number of patients involved, e.g., in the Linxian cataract study or the ARED study, beneficial effects, if evident at all, can hardly be attributed to any of the individual components of the supplement mixtures. The kind of study design may be relevant to evaluate the sense of current trends to expect miracles from supranutritional dosages of micronutrients, be they antioxidants or not. The dose-dependent switch in mechanism of action, which here has been exemplified for selenium and could easily be extended to other micronutrients, makes the outcome of such studies hard to interpret. The problem whether ocular diseases are aggravated by oxidative stress and which presumed antioxidant might retard disease manifestation could not be solved this way.

Conclusions

Selenium has an established role in ocular physiology. As an integral part of glutathione peroxidase type 1, it prevents oxidative damage and, in consequence, cataract formation in the eye lens of rodents, as is demonstrated by alimentary selenium deprivation and genetic disruption of the GPx-1 gene. The roles of other selenoproteins in the eye remain to be established.

Excess selenium that is not incorporated specifically into selenoproteins causes cataracts in postnatal animals presumably via redox cycling.

There is no theoretical basis, nor any experimental evidence for the hope that any selenium supplementation that exceeds the dietary reference intakes of 55 μg/day has a beneficial effect on the eye. Moreover, animal experimentation reveals that only severe and sustained selenium deficiency affects eye physiology. Accordingly, any reliable clinical data revealing a beneficial effect of selenium on ocular diseases are missing.

Taking into account the documented difficulties to induce any pathologically relevant selenium deficiency in the eyes of experimental animals and the known hazards of excess selenium, supranutritional selenium supplementation to prevent age-related ocular diseases can at present not be recommend.

In view of the established function of selenium in the eye, it nevertheless appears advisable to screen oxidative stress-related ocular diseases for disturbances of selenium biochemistry that, in exceptional cases, might be a cause or complicating factor of the disease.

References

1   Brune GE: Animal studies on cataract; in Taylor A (ed): Nutrition and Environmental Influences on the Eye. Boca Raton, CRC Press, 1999, pp 105–115.
2   Combs GF Jr, Combs SB: The Role of Selenium in Nutrition. Orlando, Academic Press, Harcourt Brace Jovanovich, 1986, pp 265–326, 413–461.
3   Kryukov GV, Castellano S, Novoselov SV, Lobanov AV, Zehtab O, Guigó R, Gladyshev VN: Characterization of mammalian selenoproteomes. Science 2003;300:1439–1443.
4   Birringer M, Pilawa S, Flohé L: Trends in selenium biochemistry. Nat Prod Rep 2002;19: 693–718.
5   Brigelius-Flohé R: Tissue-specific functions of individual glutathione peroxidases. Free Radic Biol Med 1999;27:951–965.
6   Schomburg L, Schweizer U, Holtmann B, Flohé L, Sendtner M, Köhrle J: Gene disruption discloses role of selenoprotein P in selenium delivery to target tissues. Biochem J 2003;370:397–402.
7   Brigelius-Flohé R, Maiorino M, Ursini F, Flohé L: Selenium: An antioxidant? in Cadenas E, Packer L (eds): Handbook of Antioxidants. New York, Dekker, 2001, pp 633–664.
8   Spallholz JE: On the nature of selenium toxicity and carcinostatic activity. Free Radic Biol Med 1994;17:45–64.
9   Flohé L, Günzler WA, Schock HH: Glutathione peroxidase: A selenoenzyme. FEBS Lett 1973;32:132–134.
10   Pirie A: Glutathione peroxidase in lens and a source of hydrogen peroxide in aqueous humour. Biochem J 1965;96:244–253.
11   Martin-Alonso JM, Ghosh S, Coca-Prados M: Cloning of the bovine plasma selenium-dependent glutathione peroxidase cDNA from the ocular ciliary epithelium: Expression of the plasma and cellular forms within the mammalian eye. J Biochem (Tokyo) 1993;114:284–291.
12   Haung W, Koralewska-Makar A, Bauer B, Akesson B: Extracellular glutathione peroxidase and ascorbic acid in aqueous humor and serum of patients operated on for cataract. Clin Chim Acta 1997;261:117–130.
13   Kinoshita JH: Selected topics in ophthalmic biochemistry. Arch Ophthalmol 1964;72:554–572.
14   Reddy DV, Kinsey VE: Studies on the crystalline lens. IX. Quantitative analysis of free amino acids and related compounds. Invest Ophthalmol 1962;1:635–641.
15   Harding JJ: Free and protein-bound glutathione in normal and cataractous human lenses. Biochem J 1970;117:957–960.

16 Srivastava SK, Beutler E: Permeability of normal and cataractous rabbit lenses to glutathione. Proc Soc Exp Biol Med 1968;127:512–514.

17 Srivastava SK, Beutler E: Cataract produced by tyrosinase and tyrosine systems in rabbits in vitro. Biochem J 1969;112:421–425.

18 Rees JR, Pirie A: Possible reactions of 1,2-naphthoquinone in the eye. Biochem J 1967;102: 853–863.

19 Pirie A, Rees JR, Holmberg NJ: Diquat cataract in the rat. Biochem J 1969;114:89P.

20 Graw J, Kratochvilova J, Summer KH: Genetical and biochemical studies of a dominant cataract mutant in mice. Exp Eye Res 1984;39:37–45.

21 Oravecz-Wilson KI, Kiel MJ, Li L, Rao DS, Saint-Dic D, Kumar PD, Provot MM, Hankenson KD, Reddy VN, Lieberman AP, Morrison SJ, Ross TS: Huntingtin interacting protein-1 mutations lead to abnormal hematopoiesis, spinal defects and cataracts. Hum Mol Genet 2004;13:851–867.

22 Reddy VN, Giblin FJ, Lin LR, Dang L, Unakar NJ, Musch DC, Boyle DL, Takemoto LJ, Ho YS, Knoernschild T, Juenemann A, Lutjen-Drecoll E: Glutathione peroxidase-1 deficiency leads to increased nuclear light scattering, membrane damage, and cataract formation in gene-knockout mice. Invest Ophthalmol Vis Sci 2001;42:3247–3255.

23 Sprinker LH, Harr JR, Newberne PM, Whanger PD, Weswig PH: Selenium deficiency lesions in rats fed vitamin E-supplemented rations. Nutr Rep Int 1971;4:335.

24 Whanger PD, Weswig PH: Effects of selenium, chromium and antioxidants on growth, eye cataracts, plasma cholesterol and blood glucose in selenium-deficient, vitamin E-supplemented rats. Nutr Rep Int 1975;12:345.

25 Chandrasekher G, Sailaja D: Alterations in lens protein tyrosine phosphorylation and phosphatidylinositol 3-kinase signaling during selenite cataract formation. Curr Eye Res 2004;28: 135–144.

26 Shearer TR, David LL, Anderson RS, Azuma M: Review of selenite cataract. Curr Eye Res 1992;11:357–369.

27 Gupta SK, Halder N, Srivastava S, Trivedi D, Joshi S, Varma SD: Green tea (*Camellia sinensis*) protects against selenite-induced oxidative stress in experimental cataractogenesis. Ophthalmic Res 2002;34:258–263.

28 Boivin P, Galand C, Bernard JF: Deficiencies in G-SH biosynthesis; in Flohé L, Benöhr HC, Sies H, Waller HD, Wendel A (eds): Glutathione. Stuttgart, Thieme, 1973, pp 146–157.

29 Beutler E, Srivastava SK: G-SH metabolism of the lens; in Flohé L, Benöhr HC, Sies H, Waller HD, Wendel A (eds): Glutathione. Stuttgart, Thieme, 1973, pp 201–205.

30 Moro F, Gorgone G, Li Volti S, Cavallaro N, Faro S, Curreri R, Mollica F: Glucose-6-phosphate dehydrogenase deficiency and incidence of cataract in Sicily. Ophthalmic Paediatr Genet 1985;5:197–200.

31 Meloni T, Carta F, Forteleoni G, Carta A, Ena F, Meloni GF: Glucose-6-phosphate dehydrogenase deficiency and cataract of patients in Northern Sardinia. Am J Ophthalmol 1990;110:661–664.

32 Bhatia RP, Patel R, Dubey B: Senile cataract and glucose-6-phosphate dehydrogenase deficiency in Indians. Trop Geogr Med 1990;42:349–351.

33 Chen Z, Zeng L, Ma Q, Su W, Mao W: The study of G6PD in erythrocyte and lens in senile and presenile cataract. Yan Ke Xue Bao 1992;8:12–5,33.

34 Assaf AA, Tabbara KF, el-Hazmi MA: Cataracts in glucose-6-phosphate dehydrogenase deficiency. Ophthalmic Paediatr Genet 1993;14:81–86.

35 Yokota T, Shiojiri T, Gotoda T, Arai H: Retinitis pigmentosa and ataxia caused by a mutation in the gene for the α-tocopherol transfer protein. N Engl J Med 1996;335:1770–1771.

36 Corrocher R, Guadagnin L, de Gironcoli M, Girelli D, Guarini P, Olivieri O, Caffi S, Stanzial AM, Ferrari S, Grigolini L: Membrane fatty acids, glutathione-peroxidase activity, and cation transport systems of erythrocytes and malondialdehyde production by platelets in Laurence Moon Barter Biedl syndrome. J Endocrinol Invest 1989;12:475–481.

37 Stanzial AM, Bonomi L, Cobbe C, Olivieri O, Girelli D, Trevisan MT, Bassi A, Ferrari S, Corrocher R: Erythrocyte and platelet fatty acids in retinitis pigmentosa. J Endocrinol Invest 1991;14:367–373.

38 Stahl W, Sies H: Antioxidant effects of carotenoids: Implication in photoprotection in humans; in Cadenas E, Packer L (eds): Handbook of Antioxidants. New York, Dekker, 2002, pp 223–233.

39   Cohen SM, Olin KL, Feuer WJ, Hjelmeland L, Keen CL, Morse LS: Low glutathione reductase
      and peroxidase activity in age-related macular degeneration. Br J Ophthalmol 1994;78:791–794.
40   Weeks DE, Conley YP, Mah TS, Paul TO, Morse L, Ngo-Chang J, Dailey JP, Ferrell RE, Gorin MB:
      A full genome scan for age-related maculopathy. Hum Mol Genet 2000;9:1329–1349.

Prof. Dr. Leopold Flohé
MOLISA GmbH, Universitätsplatz 2
DE–39106 Magdeburg (Germany)
Tel. +49 331 7480950, E-Mail l-flohe@t-online.de

Augustin A (ed): Nutrition and the Eye.
Dev Ophthalmol. Basel, Karger, 2005, vol 38, pp 103–119

....

Nutritional Supplementation to Prevent Cataract Formation

Carsten H. Meyer, Walter Sekundo

Department of Ophthalmology, Philipps University of Marburg,
Marburg, Germany

Abstract

Age-related cataract (ARC) is the leading cause of blindness in the world, particularly in developing countries. In contrast, cataract surgery has become the most frequent surgical procedure in people aged 65 years or older in the Western world, causing a considerable financial burden to the health care system. The development of cataracts is mainly an age-related phenomenon, although socioeconomic and lifestyle factors appear to influence their development, e.g. smoking has been found to directly influence ARC. A key role in the pathomechanism of the crystalline lens alteration is played by glucose metabolism and associated effected redox potential, which may induce oxidative damages. Aldose reductase blockers were able to prevent the development of diabetic cataracts in experimental studies, however clinical trials were interrupted due to unclear side effects. Other drugs with radical scavenging properties were effective in in vitro and in vivo experiments, but could not be proven to be efficient and safe in preclinical human trials. A number of epidemiological studies showed an increased risk of nuclear or cortical cataract in people with low blood levels of vitamin E. It is also known that the measured levels of ascorbic acid decline with increasing age in the lens. β-Carotin and other non-polar carotenoids seem to be missing and may therefore only play a minor role. Polarized carotenoid lutein and zeaxanthin are available in low concentrations and may therefore have some direct effects. The results of the present interventional studies are still controversial. While the Linxian studies indicated that the prevalence for nuclear cataract was reduced by the supplementation with retinol/zinc or vitamin C/molybdenum, the AREDS trial showed no effect of the antioxidant formulation on the development or progression of ARC. Again, while the REACT study demonstrated a statistically significant positive treatment effect 2 years after treatment for the US patients and for both subgroups (US + UK) after 3 years, no effect was observed in UK patients alone. In another US study, the Physician Health Study, no positive or negative effect of β-carotin was observed. Taken together, these studies suggest that any effect of antioxidants on cataract development is likely to be very small and probably is of no clinical or public health significance, thus removing a major rationale for 'anticataract' vitamin supplementation among health-conscious individuals.

Cataract Development

Age-related cataract (ARC) is the leading cause of blindness (40%) in the world [56, 71, 78]. The estimated number of currently 20 million people blinded by cataract will double by the year 2020 [3, 57, 58, 66, 69, 83]. First, this article starts with a survey about the physiology and concentration of vitamins and other substances in the lens. Subsequently, epidemiological studies on nourishment and cataract are discussed. Finally, we present several clinical interventional studies on the use of antioxidative agents and discuss their findings.

The development of cataracts is mainly an age-related phenomenon, although socioeconomic and lifestyle factors may also influence their development. Smoking has been found to directly influence ARC in many cross-sectional [3, 21, 26, 32, 38, 39, 42, 87] and longitudinal studies [57], while caffeine and drinking alone had no effect on ARC [26, 57]. In addition, both low income and educational level are also related to increased cataract rate as well as morbidity and mortality in general [15, 40, 43]. The causative pathology of these associations with increased cataracts is not clear, but may be influenced by health care, noxious environments, and high-risk behaviors [27, 46, 51, 58, 124]. Associations between cataract and risk of death were recently observed in an African-descent population [40, 70, 80].

Severe cataract is a critical problem especially in developing countries, although it also affects the more developed countries as well. Cataract surgery has become the most frequent surgical procedure in people aged ≥65 years in the USA, with an estimated annual cost of USD 3.4 billion [104]. Despite the fact that surgical treatment is relatively simple, it is known that dietary contingent aspects may play an important role in the decline of its risk, and therefore prevent or postpone costly cataract surgery. In this regard there is an ongoing discussion on the preventive effects of antioxidative and microsupplementary agents, because many oxidative, especially photooxidative processes are known to be important etiologic factors in cataract formation [10, 62, 107, 120].

The primary function of the transparent crystalline lens is to refract and focus light onto the retina [72]. As the lens does not have its own blood supply, all essential nutrients need to diffuse from the aqueous through the surrounding capsular bag and each cellular membrane into the cell. The entry metabolite transport into the epithelial cells and fibers of the lens is processed by a variety of mechanisms, including active transportation as well as diffusion processes. The maintenance of constant water hydration (approximately 69%) is also energy-consuming and depends to a great extent on metabolic activity. Significant alterations in metabolic processes result in loss of transparency and cataract formation. Most metabolic energy (90%) derives from oxidation of glucose to lactic acid by a process called anaerobic glucolysis. In addition, the lens also contains a rather unusual pathway of glucose known as the sorbitol pathway, which involves the direct utilization of non-phosphorylated sugars [30].

As the lens ages, there is a progressive increase in the level of albuminoid proteins, which render more of the lens fibers. The structural proteins (α-, β-, and γ-crystallines) constitute most of the dry weight of the lens. During aging the concentration of various lens proteins increases, especially albuminoid fraction α-crystallines are the largest (800 kDa, compared to 20 kDA for the γ-crystalline) [59, 60, 81, 82].

The general pathogenesis of human cataracts is believed to be the result of multiple factors acting over many years. Mechanisms of syn- and co-cataracto-genesis [16, 26, 33] explain cataract formation due to an accumulation of several cataract risk factors. Syn-cataractogenesis represents the combination of two (or more) damaging factors that lead in combination to lens opacities. In co-cataractogenesis the direct cataractogenic effect of a substance is promoted when it is in combination with a subliminal factor that, on its own, has no effect. The term *age-related cataract* includes three different forms of lens opacity: cortical, nuclear, and posterior subcapsular (PSC) (fig. 1–3). Each of these has its own distinctive pathogenic changes and distinctive risk factors. While ultraviolet light exposure [100, 111, 122] and steroid administration have been linked to the cortical cataract, cigarette smoking has been consistently associated with the nuclear cataract [65, 67, 110]. A variety of biochemical, animal, and human studies suggested that oxidative changes of the lens proteins may cause lenses to become opaque [7, 9, 11, 14, 17, 119, 126].

Early investigations in the 1930s on the pathophysiology of cataract formation led to considerations that the metabolisms can be influenced by medical treatment [85]. H.K. Müller, Chairman at the Department of Ophthalmology at the University of Bonn, pursued a conservative cataract therapy and established the Clinical Institute for Experimental Ophthalmology [85, 86, 88] some 50 years ago. Major investigations during the 1950s to 1970s on lens metabolism, aging processes of the lens proteins and their impact on lens transparency were the key results to consider a therapy for 'senile cataract' [3, 44, 45, 50, 89].

A key role in the pathomechanism of the crystalline lens alteration is played by the glucose metabolism and associated effected redox potential, which may induce oxidative damages [48, 89, 93]. Therefore, aging is not the primary cause, but rather a background of the cataractogenesis as some interactions occur during life. Multiple cross-reactions are responsible for the multifactorial cataractogenesis [100, 111, 122]. With increasing light absorption,

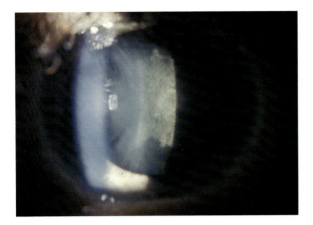

Fig. 1. An anterior cortical cataract is seen as a diffuse whitish opacity virtually under the anterior lens capsule. It can be located centrally (as in this photograph), close to the equator of the lens or 'spread' from the periphery toward the optical center.

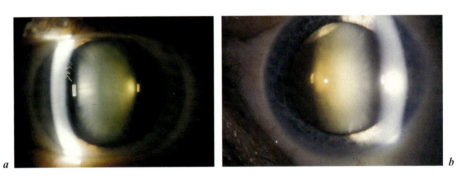

a   *b*

Fig. 2. *a* A typical nuclear cataract (2+ nuclear sclerosis) is characterized by a yellowish appearance of the nucleus when viewed at the slitlamp. *b* In advanced stages of nuclear sclerosis the lens turns brown: an entity known as 'brunescent cataract'.

the crystalline lens builds additional disulfide bridges and glycosylations at the protein level exaggerating to large biochemical complexes and networks. It is noteworthy to stress that diabetes may also trigger certain cataractogenic interactions. Two additional mechanisms are also briefly mentioned, namely (a) the damage of epithelial cells before or during their differentiation may induce a false differentiation or persisting cellular damage, leading to PSC cataract, and (b) the entire metabolism of the lens may be disturbed by toxic events. All three conditions mentioned cause oxidative damages and persistent impairment. The above-named results led to a variety of experimental and

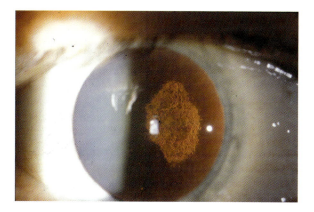

Fig. 3. A PSC cataract usually presents as a large white plaque behind the nucleus and is best appreciated using retroillumination.

therapeutical attempts and trials on cataract prophylaxis, which can be divided essentially into two groups.

Therapeutics

Group A: Substances and substance mixtures aimed at slowing down the aging process of the lens through an optimization of the lenses provision.

These preparations mainly target the diabetic metabolism (type 1 diabetes mellitus), its effect on the activity of aldose reductase and the resulting accumulation of sorbitol. Efficient blockers of aldose reductase were developed in the 1960s and 1970s [89].

Aldose reductase blockers such as Sorbinil® (Pfizer) were able to prevent the development of diabetic cataracts in experimental studies. Although systemically or locally given, they could prevent cataract in animal experiments [5]; clinical trials using Sorbinil® were interrupted due to unclear side effects. In clinical trials other products demonstrated an extreme long persistence in the lens, therefore no suitable dosage could be found and the clinical trials were stopped. Pyruvate was also used to prevent oxidative damage of the crystalline lens [29]. Although promising attempts to stimulate the metabolism of glutathione were published in experimental studies [9, 18, 23], there are no clinical studies investigating initial drugs (prodrug) products.

Finally, Tempol® was developed to protect lens epithelial cells from H_2O_2 radical damages [92]. Tempol® was effective in vitro and also in vivo, preventing lens damage in rabbits after X-ray exposure [97]. However, none of these experimental agents could be proven to be efficient and safe in preclinical or human trials.

A variety of different agents have been developed in experimental investigations to prevent specific entities of cataract formations, however only a minority reached clinical trials. None of the tested drugs demonstrated a convincing result supporting their frequent and regular use.

Group B: The second group targets protein synthesis, vitamin supplementation and protection from oxidative damages.

Nutritional factors, particularly vitamins with antioxidant properties, have been implicated as playing a protective role in the development and progression of cataract [1]. Vitamin A, niacin, riboflavin, thiamin, folate, and vitamin B_{12} all appeared to be protective, either individually or as constituents of multivitamin preparations.

Oxidative stress in the ocular lens seems to play an important role in the etiology of cataracts as a structural change of lens proteins may induce lens opacification [8, 61]. In humans, lens opacities are associated with oxidative protein modifications, leading to aggregation and loss of transparency. The risk of cataract formation can be decreased by vitamins with antioxidant properties in order to ameliorate the risk of oxidative stresses [9, 11, 17, 101, 119, 126]. Animal studies demonstrated a protective effect of different vitamins. Cross-sectional studies suggested that dietary vitamin supplement use may delay the onset of both nuclear and cortical cataract and could be associated with a reduced prevalence of the latter [7, 22, 23, 29, 34].

Vitamin C

The eye is a specialized organ, which can be damaged by a variety of agents and stimuli. Most damages have their origin in oxidative processes. Numerous in vitro and in vivo experimental animal studies have demonstrated that antioxidative agents can reinforce intrinsic antioxidative defense mechanisms within the eye [30]. The concentration of vitamin C (ascorbic acid) is higher in the anterior chamber aqueous and the lens of the eye, compared to the plasma level in the entire body [19]. Taylor et al. [109] investigated the relation between the alimentary uptake of vitamin C and its concentration in blood plasma, as well as in the anterior chamber and lens. They demonstrated that plasma and anterior chamber aqueous were saturated by receiving a supplementation of <250 mg ascorbic acid daily. However, there was no saturation in the lens itself even after a supply of 1.5 g vitamin C per day. On the other hand it is also known that the measured levels of ascorbic acid decline with increasing age in the lens [114] and in the anterior chamber [21], whereas lower levels correlated with increasing cataract progression [11, 27]. Reiss et al. [94] reported higher ascorbic acid concentrations in the anterior chamber of day-active animals, whereas night-active animals had a much lower concentration. This supports the hypothesis that water-soluble antioxidative agents like vitamin C may play a role in ocular light protection [7, 36, 48, 95, 115, 117, 118].

Vitamin E

Tocopherol (vitamin E) is the primary lipid-soluble antioxidant in biological membrane stabilizers [7, 12]. Numerous studies have found benefits arising from the use of vitamin E [36, 48, 71, 79, 95, 99, 113, 116]. It plays an important role in maintaining membrane integrity and glutathione (GSH)-1 synthesis cycling [1]. Both vitamin E and GSH exhibit a close interaction in their antioxidant function [9]. A number of epidemiological studies showed an increased risk of nuclear or cortical cataract in people with low blood levels of vitamin E [12]. The deficiency of vitamin E may lead to the formation of toxic peroxides and malondialdehydes due to an overall acceleration of tissue oxidation resulting in lens opacification. The dietary amount of vitamin E supplementation clearly effects the alteration of protein structure accompanied by the change of glutathione level in the lens. Elevation of GSH level in the lens might effectively defend lens membrane from lipid peroxidation, which in turn probably reduces the risk of protein oxidation. Accordingly, adequate antioxidant supplements such as vitamin E may protect lens proteins from oxidative insults. On the other hand, an excess of vitamin E intake may reduce its intrinsic advantage of antioxidation.

Carotenoids

Several lipophilic agents other than vitamin E, e.g. carotenoids, lutein and zeaxanthin, have a known antioxidative potential for the eye [14, 34, 99]. Although they can all be within the lens, the dominating substance remains vitamin E, whereas lutein and zeaxanthin have lower levels in the lens and are more important agents in the retina [8]. The concentration of vitamins and carotenoids in cataract lenses is higher in the outer layers including the cortex and epithelial cells [31]. Bates and Cowen [13] analyzed the concentration of vitamin E, and several carotenoids including β-cryptoxanthin, α-carotin and β-carotin in human lenses. They confirmed the dominance of vitamin E, whereas the tested quantities of carotenoid were much lower, even lower than that of lutein and zeaxanthin. Patients who received 18 mg β-carotin daily for 3 months prior to their cataract surgery had no significant difference for β-carotin concentration in the lens nucleus.

In summary, vitamins C and E are the most important antioxidative agents for the lens. β-Carotin and other nonpolar carotenoids seem to be missing and may therefore only play a minor role. Polarized carotenoid lutein and zeaxanthin are available in low concentrations and may therefore have some direct effects.

Epidemiological Studies

Several epidemiological studies correlated the blood plasma concentrations of micro-agents with the presence or absence of cataract formation [41, 91, 96]. The comparison of different epidemiological studies is very difficult, as the design (retrospective, prospective, etc.) as well as the classification schemes for cataract may differ between different study groups. It is not surprising therefore that the epidemiological evidence does not appear to be in total agreement [64, 84, 102]. However, many of these studies indicate that vitamin C, vitamin E and carotenoids may contribute to a decline of the ARC risk. Although theses studies cannot prove a causal connection, they may give valuable statistical correlations between nutrition and cataract formation and their possible impact on ARCs [1–4, 51–58, 63, 83, 105, 106].

There are numerous survey articles on the influences of nourishment-dependent factors such as the intake dosage, the present plasma concentration of antioxidative agents and the ARC risk. The most current survey was published by Taylor and Hobbs [108] who demonstrated favorable effects on ARC risk with a low supply of 250 mg vitamin C, 90 mg vitamin E and 4 mg lutein daily, without toxic side effects. Mares-Perlman et al. [73, 74] recently summarized evidence for carotenoid, lutein and zeaxanthin to exert a positive effect with regard to cataract development.

Clinical Interventional Studies

Evidence of effectiveness (or the lack of it) may, in contrast to epidemiological studies, be more defined in clinical interventional studies [18, 20, 21, 37, 48, 49, 63, 68, 71, 73, 74, 125]. However, in cataracts, it may be difficult to select the ideal design, definition of clinical functional parameters and final points, as they have a very slow progress and may already start during early childhood. On the other hand, longer interventional studies are associated with additional considerable problems, including patient dropouts or compliance to receipt-prescribed nourishment supplementation. Nevertheless, some clinical interventional studies were accomplished in which the cataract risk was judged with participants who had received placebo or antioxidative agents. However, in some of these studies the primary goal was not to evaluate the formation of cataract, but rather of cancer (Linxian and ATBC) or cardiovascular events

(PHS). These studies evaluated the cataract formation only as a secondary outcome measure in a subgroup analysis of the total study population.

The following section describes three of these general studies and in addition the AREDS trial aligned on age-related eye diseases and the REACT study investigating the cataract formation exclusively.

'Linxian' Studies

The Linxian studies are the only completed randomized trial on vitamin supplements and cataract, and it found that some vitamins may have a protective role in cataract development [19, 103]. The studies included almost 4,000 participants aged 45–74 years in the rural communities of Linxian, China. In the first study the participants received either a multivitamin/mineral preparation or placebo. In the second study, a more complex factorial design was used to determine the effects of four different vitamins/mineral material combinations: (a) retinol (5,000 IU) and zinc (22 mg); (b) riboflavin (3 mg) and niacin (40 mg); (c) vitamin C (120 mg) and molybdenum (30 mg), and (d) vitamin E (30 mg), β-carotin (15 mg) and selenium (50 mg).

Eye investigations were carried out at the end of the 5- to 6-year follow-up period in order to determine cataract prevalence. In the first study in patients aged 65–74 years who had received the multivitamin/mineral supplementation, a statistically significant reduced nuclear cataract prevalence of 43% was determined. In the second study, the riboflavin/niacin preparation was associated with a lower prevalence of nuclear cataract, but also with a higher risk of PSC cataract. The results also disclosed that the prevalence for nuclear cataract was reduced by the supplementation with retinol/zinc or vitamin C/molybdenum. The major weakness of the Linxian studies was the lack of ophthalmologic examination at the beginning of the study.

The AREDS Trial

The Age-Related Eye Disease Study (AREDS) [1–4], designed as a prospective, controlled study and including 4,629 participants aged 55–80 years, tested whether a daily supplementation with high-dose antioxidant combination, consisting of vitamin E (400 IU daily), vitamin C (500 mg daily), and β-carotene (15 mg daily) with and without zinc (80 mg zinc oxide) and copper (2 mg copper II oxide) or placebo for an average duration of 6.3 years, influences the development and progression of ARC. The report was a direct extension of the previously published Blue Mountains Eye Study report, conducted during the period 1992–1994, on the relationship between the three principal types of cataract and a wide range of dietary macro- and micronutrients, including vitamins [9]. The cataract-related outcome measure was based on combined data of appearance or progression of any of three main cataract types and lens extraction, referred to as a 'cataract event'. No effect of the antioxidant formulation was observed on the development or progression of any of the three types of age-related lens opacities or cataract surgery. The results contrast with other observational data that has linked the intake of antioxidant vitamins with a reduced incidence of cataract. It also contrasts with in vivo and in vitro data attributing the development of senile cataract to oxidative injury of lens proteins and lipids. However, no undesirable effects were observed and the findings are in keeping with recent cardiovascular studies in which observational epidemiological analyses had identified a protective effect of antioxidant vitamins.

The REACT Study

The Roche European-American Cataract Trial (REACT) [25] was a prospective, randomized, placebo-controlled, double-blind, multicenter study performed in the UK and USA. Subjects received a combination of antioxidative agents (vitamin C, 750 mg; vitamin E, 600 IU, and β-carotin, 18 mg, daily for 3 years). 445 patients were recruited and 297 randomized for the study; 231 (78%) patients were observed for 2 years and 158 (53%) for 3 years. A statistically significant positive treatment effect appeared 2 years after treatment in the US patients and for both subgroups (US + UK) after 3 years. However, there was no statistically significant treatment effect in the UK group at any observational point.

The PHS Studies

The PHS (Physicians' Health Study) was a controlled, clinical interventional study including about 22,000 male subjects in the USA, who received daily either 50 mg β-carotin or low-dose aspirin [24, 90, 98]. Approximately 21,000 participants, who had no cataracts at the beginning of the study, underwent an eye examination. After a median duration of 13.2 years, 2,017 cataracts were confirmed, 998 in the β-carotin-group and 1,019 in the placebo group. There was no positive or negative effect of β-carotin on cataract prevention. The currently ongoing PHS-II study will be carried out to examine cataract as well as cancer and cardiovascular diseases in a follow-up study.

The ATBC Study

The α-tocopherol β-carotene (ATBC) study [6] was a controlled, clinical study conducted in Finland including 29,133 male subjects aged 50–69 years who were strong smokers. The risk of developing lung cancer was investigated when the patients were treated with α-tocopherol (50 mg/day) and or β-carotin (20 mg/day) over 5–8 years. At the end of this study, a randomly selected group of 1,828 participants received an eye examination according to the LOCS II criteria to determine the presence of cortical, nuclear or subcapsular cataract [6].

No statistically significant differences could be assessed between each group. This result confirmed the zero hypothesis for β-carotin (no effect, no risk) similar to the previously mentioned PHS study. In addition, the selected vitamins E dose (50 mg) may have been too low in order to determine an effect on cataract formation in the examined study population.

The VECAT Study

The Vitamin E, Cataract and Age-Related Maculopathy Trial (VECAT) [35] was a controlled clinical study carried out in Australia, investigating whether the supplementation with vitamin E (500 mg/days) over 4 years reduces the incidence or progression of cataract and age-related macular degeneration. 1,204 subjects at the age of 55–80 years were included in the study. Lens opacifications were documented using an EAS-1000 camera (Nidek Inc.) and graded according to the Wilmer clinical schemes for cataract. The early results of the VECAT study demonstrated that pharmacological doses of vitamin E had no clinically significant effect on either incidence or progression of nuclear or cortical cataract. The groups did not differ in the proportions undergoing surgery for cataract removal. Although a notable difference in incidence of new PSC cataract between the two treatment groups was found, conclusions about the significance of this finding were limited by low statistical power for this type of cataract. The results are likely to be widely extended to elderly populations with high proportions of non-smokers, and the observed cataract progression rates are broadly in keeping with similar populations studied elsewhere [22]. The results of VECAT are similar to those of the AREDS group [1–4].

Conclusion

Much disagreement exists over which vitamins protect against cataract. However, numerous studies have found some benefits arising from the use of carotenoids [23, 37, 106], vitamin A [22, 23, 27, 46, 73, 106], vitamin C [37, 48, 73, 65, 95, 125], vitamin E [34, 60, 65, 68, 123], multivitamins [11, 12, 16, 19–21], and niacin, thiamin, or riboflavin [9, 11, 12, 18]. They are in keeping with the published results of the AREDS trial, in which an antioxidant combination failed to influence cataract development over an average 6.3-year interventional period. The power of the latter study was sufficient to exclude more than 13% reduction in the odds of any lens event [1–4]. Taken together, these studies suggest that any effect of antioxidants on cataract development is likely to be very small and probably is of no clinical or public health significance, thus removing a major rationale for vitamin E supplementation among health-conscious individuals. At present though, this substance is used extensively, primarily by individuals who anticipate benefits in terms of reduced coronary heart disease and cataract, as well as a slowing of the aging process.

None of the described medical prevention studies demonstrated a convincing success. The results suggest that the protective effect of antioxidant vitamins identified in earlier case-control and epidemiological studies may have been confounded by other lifestyle factors [58]. Individuals who are regular users of multivitamin preparations are more likely to be health conscious. Further randomized clinical trials and/or meta-analysis of data from previous randomized trials will be needed to determine whether cataract could be prevented or delayed by using vitamin supplements of any type. Critical evaluation of potential side effects [116], particularly using high-dose antioxidative agents, is mandatory in view of the fact that most agents can neither prevent nor prolong the progression of the lens opacification. In conclusion, there is currently no drug on the market that can prevent ARC.

References

1   AREDS Research Group: The Age-Related Eye Disease Study (AREDS). Design implications. AREDS Report No 1. Control Clin Trials 1999;20:573–600.
2   AREDS Research Group: The Age-Related Eye Disease Study (AREDS) system for classifying cataracts from photographs: AREDS Report No 4. Am J Ophthalmol 2001;131:167–175.
3   AREDS Research Group: Risk factors associated with age-related nuclear and cortical cataract: A case-control study in the Age-Related Eye Disease Study, AREDS Report No 5. Ophthalmology 2001;108:1400–1408.
4   AREDS Research Group: A randomized, placebo-controlled, clinical trial of high-dose supplementation with vitamins C and E and β-carotene for age-related cataract and vision loss: AREDS Report No 9. Arch Ophthalmol 2001;119:1439–1452.
5   Ajiboye R, Harding JJ: The non-enzymic glycosylation of bovine lens proteins by glycosamine and its inhibition by aspirin, ibuprofen and glutathione. Exp Eye Res 1989;49:31–41.
6   ATBC Study Group: The effect of vitamin E and β-carotene on the incidence of lung cancer and other cancers in male smokers. N Engl J Med 1994;330:1029–1035.
7   Augustin A, Spitznas M, Breipohl W, Wegener A: Evidence for the prevention of oxidative tissue damage in the inner eye by vitamin E and vitamin C. Ger J Ophthalmol 1992;1:394–398.
8   Augustin AJ, Böker SHH, Blumenröder, Lutz J, Spitznas M: Free radical scavenging and antioxidant activity of allopurinol and oxypurinol in experimental lens-induced uveitis. Invest Ophthalmol Vis Sci 1994;35:3897–3904.
9   Augustin A, Dick HB, Winkgen A, Schmidt-Erfurth U: Ursache und Prävention oxidativer Schäden des Auges – eine aktuelle Bestandesaufnahme. Ophthalmologe 2001;98:776–796.
10 Augustin AJ: Kataraktprävention. Prophylaxe mit Nebenwirkung? Ophthalmologe 2003;100:175.
11 Babizhayev MA: Failure to withstand oxidative stress induced by phospholipid hydroperoxides as a possible cause of the lens opacities in systemic diseases and ageing. Biochim Biophys Acta 1996;1315:87–99.
12 Bates CJ, Chen SJ, MacDonald A, Holden R: Quantitation of vitamin E and a carotenoid pigment in cataractous human lenses, and the effect of a dietary supplement. Int J Vitam Nutr Res 1996; 66:316–21.
13 Bates CJ, Cowen TD: Effects of age and dietary vitamin C on the contents of ascorbic acid and acid-soluble thiol in lens and aqueous humour of guinea-pigs. Exp Eye Res 1988;46:937–945.

14 Bernstein PS, Khachik F, Carvalho LS, Muir GJ, Zhao DY, Katz NB: Identification and quantitation of carotenoids and their metabolites in the tissues of the human eye. Exp Eye Res 2001;72:215–223.

15 Benson WH, Farber ME, Caplan RJ: Increased mortality rates after cataract surgery. A statistical analysis. Ophthalmology 1988;95:1288–1292.

16 Bhuyan KC, Bhuyan DK: Molecular mechanisms of cataractogenesis. III. Toxic metabolites of oxygen as initiators of lipid peroxidation and cataract. Curr Eye Res 1984;3:67–81.

17 Boscia F, Grattagliano I, Vendemiale G, et al: Protein oxidation and lens opacity in humans. Invest Ophthalmol Vis Sci 2000;41:2461–2465.

18 Brown NA, Bron AJ, Harding JJ, Dewar HM: Nutrition supplements and the eye. Eye 1998;12: 127–133.

19 Bunce GE: Evaluation of the impact of nutrition intervention on cataract prevalence in China. Nutr Rev 1994;52:99–101.

20 Bunce GE, Kinoshita J, Horwitz J: Nutritional factors in cataract. Annu Rev Nutr 1990;10:233–254.

21 Chasan-Taber L, Willett WC, Seddon JM, et al: A prospective study of vitamin supplement intake and cataract extraction among US women. Epidemiology 1999;10:679–684.

22 Christen WG: Antioxidants and eye disease. Am J Med 1994;97:7S–14S.

23 Christen WG: Antioxidant vitamins and age-related eye disease. Proc Assoc Am Physicians 1999;111:16–21.

24 Christen WG, Gaziano JM, Hennekens CH: Design of Physicians' Health Study II – A randomized trial of β-carotene, vitamins E and C, and multivitamins, in prevention of cancer, cardiovascular disease, and eye disease, and review of results of completed trials. Ann Epidemiol 2000; 10:125–134.

25 Chylack LT, Brown NP, Bron A, Hurst M, Kopcke W, Thien U, Schalch W: The Roche European-American Cataract Trial (REACT): A randomized clinical trial to investigate the efficacy of an oral antioxidant micronutrient mixture to slow progression of age-related cataract. Ophthalmic Epidemiol 2002;9:49–80.

26 Cumming RG, Mitchell P: Alcohol, smoking, and cataracts: The Blue Mountains Eye Study. Arch Ophthalmol 1997;115:1296–1303.

27 Cumming RG, Mitchell P, Smith W: Diet and cataract: The Blue Mountains Eye Study. Ophthalmology 2000;107:450–456.

28 Delcourt C, Cristol JPC, Leger CL, Descomps B, et al: Association of antioxidant enzymes with cataract and age-related macular degeneration. The POLA Study. Ophthalmology 1999;106:215–222.

29 Delcourt C, Carriere I, Delage M, Descomps B, et al: Association of cataract with antioxidant enzymes and other risk factors. Ophthalmology 2003;110:2318–2326.

30 Duncan G, Wormstone IM, Davies PD: The aging human lens: Structure, growth, and physiological behaviour. Br J Ophthalmol 1997;81:818–823.

31 Fecondo JV, Augusteyn RC: Superoxide dismutase, catalase and glutathione peroxidase in the human cataractous lens. Exp Eye Res 1983;36:15–23.

32 Flaye DE, Sullivan KN, Cullinan TR, et al: Cataracts and cigarette smoking. The City Eye Study. Eye 1989;3:379–384.

33 Fu S, Dean R, Southan M, Truscott R: The hydroxyl radical in lens nuclear cataractogenesis. J Biol Chem 1998;273:28603–28609.

34 Gale CR, Hal NFl, Phillips DI, Marty CN: Plasma antioxidant vitamins and carotenoids and age-related cataract. Ophthalmology 2001;108:1992–1998.

35 Garrett SK, McNeil JJ, Silagy C, et al: Methodology of the VECAT study: Vitamin E intervention in cataract and age-related maculopathy. Ophthalmic Epidemiol 1999;6:195–208.

36 Gey KF: Vitamins E plus C and interacting conutrients required for optimal health. A critical and constructive review of epidemiology and supplementation data regarding cardiovascular disease and cancer. Biofactors 1998;7:113–174.

37 Hankinson SE, Stampfer MJ, Seddon JM, et al: Nutrient intake and cataract extraction in women: A prospective study. BMJ 1992;305:335–339.

38 Harding JJ, van Heyninggen R: Beer, cigarettes, and military work as risk factors for cataract. Dev Ophthalmol 1989;17:13–16.

39 Harding JJ: Cigarettes and cataract: Cadmium or lack of vitamin C? Br J Ophthalmol 1995;79: 199–201.

40   Hennis A, Wu SY, Li X, Nemesure B, Leske MS: Lens opacities and mortality. Ophthalmology 2001;108:498–504.

41   Hiller R, Sperduto RD, Ederer F: Epidemiologic associations with nuclear, cortical, and posterior subcapsular cataracts. Am J Epidemiol 1986;124:916–925.

42   Hiller R, Sperduto RD, Podgor MJ, et al: Cigarette smoking and the risk of development of lens opacities. The Framingham studies. Arch Ophthalmol 1997;115:1113–1118.

43   Hirsch RP, Schwartz B: Increased mortality among elderly patients undergoing cataract extraction. Arch Ophthalmol 1983;101:1034–1037.

44   Hockwin O, Korte I, Noll E, Heiden M, Konopka R, Hagenah I, Hurtado R: Is it possible to maintain a normal glutathione level in lenses in vitro? Graefes Arch Clin Exp Ophthalmol 1985;222: 142–146.

45   Hockwin O, Schmitt C: Stellenwert der Antikataraktika. Fortschr Ophthalmol 1990;87:9–13.

46   Hodge WG, Whitcher JP, Satariano W: Risk factors for age-related cataracts. Epidemiol Rev 1995;17:336–346.

47   Holleschau AM, Rathbun WB, Nagasawa HAT: A HPLC radiotracer method for assessing the ability of *L*-cysteine prodrugs to maintain glutathione levels in the cultured rat lens. Curr Eye Res 1996;15:501–510.

48   Jacques PF, Taylor A, Hankinson SE, et al: Long-term vitamin C supplement use and prevalence of early age-related lens opacities. Am J Clin Nutr 1997;66:911–916.

49   Jacques PF, Chylack LT Jr, Hankinson SE, et al: Long-term nutrient intake and early age-related nuclear lens opacities. Arch Ophthalmol 2001;119:1009–1019.

50   Kador PF: Overview of the current attempts toward the medical treatment of cataract. Ophthalmology 1983;90:352–364.

51   Klein R, Klein BEK, Jensen SC, Moss SE, Cruickshanks KJ: The relation of socioeconomic factors to age-related cataract, maculopathy, and impaired vision. The Beaver Dam Eye Study. Ophthalmology 1994;101:1969–1979.

52   Klein R, Wang Q, Klein BEK, et al: The relationship of age-related maculopathy, cataract, and glaucoma to visual acuity. Invest Ophthalmol Vis Sci 1995;36:182–191.

53   Klein R, Klein BEK, Moss SE: Age-related eye disease and survival. The Beaver Dam Eye Study. Arch Ophthalmol 1995;113:333–339.

54   Klein BEK, Klein R: Cataracts and macular degeneration in older Americans. Arch Ophthalmol 1982;100:571–573.

55   Klein BEK, Klein R, Linton KLP: Prevalence of age-related lens opacities in a population. The Beaver Dam Eye Study. Ophthalmology 1992;99:546–552.

56   Klein BEK, Klein R, Moss SE: Incident cataract surgery. The Beaver Dam Eye Study. Ophthalmology 1997;104:573–580.

57   Klein BEK, Klein R, Lee KE: Incidence of age-related cataract. The Beaver Dam Eye Study. Arch Ophthalmol 1998;116:219–225.

58   Klein BEK, Klein R, Lee KE: Incident cataract after a five-year interval and lifestyle factors: The Beaver Dam Eye Study. Ophthalmic Epidemiol 1999;6:247–255.

59   Kinoshita J: Mechanisms initiating cataract formation. Invest Ophthal Vis Sci 1974;13:713–724.

60   Knekt P, Heliovaara M, Rissanen A, et al: Serum antioxidant vitamins and risk of cataract. BMJ 1992;305:1392–1394.

61   Koch FHJ, Augustin AJ, Grus FH, Spitznas M: Effects of antioxidants on lens-induced uveitis. Ger J Ophthalmol 1996;5:185–188.

62   Kottler UB, Dick HB, Augustin AJ: Ist Katarakt vermeidbar? Ophthalmologe 2003;100: 190–196.

63   Kuzniarz M, Mitchell P, Cumming RG, Flood VM: Use of vitamin supplements and cataract: The Blue Mountains Study. Am J Ophthalmol 2001;132:19–26.

64   Leibowitz HM, Krueger DE, Maunder LR, et al: The Framingham Eye Study monograph: An ophthalmological and epidemiological study of cataract, glaucoma, diabetic retinopathy, macular degeneration, and visual acuity in a general population of 2,631 adults, 1973–1975. Surv Ophthalmol 1980;24(suppl):335–610.

65   Leske MC, Chylack LT Jr, Wu SY: The Lens Opacities Case-Control Study. Risk factors for cataract. Arch Ophthalmol 1991;109:244–251.

66   Leske MC, Wu SY, Hyman L, et al: Biochemical factors in the lens opacities. Case-control study. The Lens Opacities Case-Control Study Group. Arch Ophthalmol 1995;113:1113–1119.

67   Leske MC, Connell AMS, Wu SY, et al: Prevalence of lens opacities in the Barbados Eye Study. Arch Ophthalmol 1997;115:105–111.

68   Leske MC, Chylack LT, He Q, et al: Antioxidant vitamins and nuclear opacities: The Longitudinal Study of Cataract. Ophthalmology 1998;105:831–836.

69   Leske MC, Wu SY, Nemesure B, et al: Incidence and progression of lens opacities in the Barbados Eye Studies. Ophthalmology 2000;107:1267–1273.

70   Leske MS, Wu SY, Nemesure B, Hennis A: Risk factors for incident nuclear opacities. Ophthalmology 2002;109:1303–1308.

71   Lonn E, Yusuf S, Dzavik V, et al: SECURE Investigators. Effects of ramipril and vitamin E on atherosclerosis: The study to evaluate carotid ultrasound changes in patients treated with ramipril and vitamin E (SECURE). Circulation 2001;103:919–925.

72   Lutz J: Physiologie und Biochemie der Linse; in Augustin AJ (ed): Augenheilkunde, ed 2. Berlin, Springer, 2001, pp 1127–1130.

73   Mares-Perlman JA, Klein BE, Klein R, Ritter LL: Relation between lens opacities and vitamin and mineral supplement use. Ophthalmology 1994;101:315–325.

74   Mares-Perlman JA, Brady WE, Klein BE, et al: Diet and nuclear lens opacities. Am J Epidemiol 1995;141:322–334.

75   Mares-Perlman JA: Too soon for lutein supplements. Am J Clin Nutr 1999;70:431–432.

76   Mares-Perlman JA, Lyle BJ, Klein R, Fisher AI, Brady WE, Langenberg GM, Trabulsi JN, Palta M: Vitamin supplement use and incident cataracts in a population-based study. Arch Ophthalmol 2000;118:1556–1563.

77   Mares-Perlman JA, Millen AE, Ficek TL, Hankinson SE: The body of evidence to support a protective role for lutein and zeaxanthin in delaying chronic disease. Overview. J Nutr 2002;132:518–524.

78   McCarty CA, Mukesh BN, Dimitrov PN, Taylor HR: The incidence and progression of cataract in the Melbourne Visual Impairment Project. Am J Ophthalmol 2003;136:10–17.

79   McNeil JJ, Robman L, Tikellis G, Sinclair MI, McCarty CA, Taylor HR: Vitamin E supplementation and cataract: Randomized controlled trial. Ophthalmology 2004;111:75–84.

80   Meddings DR, Marion SA, Barer ML, et al: Mortality rates after cataract extraction. Epidemiology 1999;10:288–293.

81   Meyer CH, Abele B, Soergel F, Pechhold W, Laqua H: Die viscoelastischen Eigenschaften von humanen Linsenkernen verschiedenen Alters gemessen mit der dynamisch-mechanischen Analyse (DMA); in Ohrloff, et al (eds): 11. Kongress der DGII. Berlin, Springer, 1997, vol 11, pp 460–465.

82   Meyer CH, Soergel F: Akkommodationsverlust als Zeichen veränderter biomechanischer Eigenschaften der Linse. Mitt Rhein Westfäl Augenärzte 1998;25:156–159.

83   Mitchell P, Cumming RG, Attebo K, Panchapakesan J: Prevalence of cataract in Australia: The Blue Mountains Eye Study. Ophthalmology 1997;104:581–588.

84   Mohan M, Sperduto RD, Angra SK, et al: India-US case-control study of age-related cataracts. India-US Case-Control Study Group. Arch Ophthalmol 1989;107:670–676.

85   Müller HK: Zur Kritik von A. Vogt an meinem Vortrag 'Über die Altersstargenese' in Heidelberg. Klin Monatsbl Augenheilkd 1939;102:378–383.

86   Müller HK, Buschke W: Vitamin C in Linse, Kammerwasser und Blut bei normalem und pathologischem Linsenstoffwechsel. Arch Augenheilkd 1934;108:368–390.

87   Munoz B, Tajchman U, Bochow T, West S: Alcohol use and risk of posterior subcapsular opacities. Arch Ophthalmol 1993;111:110–112.

88   Ohrloff C, Hockwin O, Olson R, Dickmann S: Glutathione peroxidase, glutathione reductase and superoxide dismutase in the aging lens. Curr Eye Res 1984;3:109–115.

89   Ohrloff C, Hockwin O, Korte I, Wegener A: Aldose-Reduktase-Hemmer – ein neuer Weg zur Vermeidung diabetischer Linsenveränderungen? Klin Monatsbl Augenheilkd 1986;189:361–362.

90   PHS Group: Final report on the aspirin component of the ongoing Physicians' Health Study. N Engl J Med 1989;321:129–135.

91   Podgor MJ, Cassel GH, Kannel WB: Lens changes and survival in a population-based study. N Engl J Med 1985;313:1438–1444.

92 Reddan JR, Sevilla MD, Giblin FJ, et al: The superoxide dismutase mimic Tempol® protects cultured rabbit lens epithelial cells from hydrogen peroxide insult. Exp Eye Res 1993;56:543–554.

93 Reddan JR, Giblin FJ, Kadry R, Leverenz V, Dziedzic DC: Protection from oxidative insult in glutathione-depleted lens epithelial cells. Exp Eye Res 1999;68:117–127.

94 Reiss GR, Werness PG, Zollman PE, Brubaker RF: Ascorbic acid levels in the aqueous humor of nocturnal and diurnal mammals. Arch Ophthalmol 1986;104:753–755.

95 Robertson JM, Donner AP, Trevithick JR: A possible role for vitamins C and E in cataract prevention. Am J Clin Nutr 1991;53:346S–351S.

96 Schalch W, Chylack LT: Antioxidative Mikronährstoffe und Katarakt. Ophthalmologe 2003;100: 181–189.

97 Sasaki H, Lin LR, Yokoyama T, Sevilla MD, Reddy VN, Giblin FJ: Tempol® protects against lens DNA strand breaks and cataract in the X-rayed rabbit. Invest Ophthal Vis Sci 1998;39:544–552.

98 Seddon JM, Christen WG, Manson JE, et al: The use of vitamin supplements and the risk of cataract among US male physicians. Am J Public Health 1994;84:788–792.

99 Shih S, Weng YM, Chen S, et al: FT-Raman spectroscopic investigation of lens proteins of tilapia treated with dietary vitamin E. Arch Biochem Biophys 2003;420:79–86.

100 Skalka HW, Prchal JT: Effect of corticosteroids on cataract formation. Arch Ophthalmol 1980;98:1773–1777.

101 Spector A: Oxidative stress-induced cataract: Mechanism of action. FASEB J 1995;9:1173–1182.

102 Sperduto RD, Seigel D: Senile lens and senile macular changes in a population-based sample. Am J Ophthalmol 1980;90:86–91.

103 Sperduto RD, Hu TS, Milton RC, et al: The Linxian cataract studies. Two nutrition intervention trials. Arch Ophthalmol 1993;111:1246–1253.

104 Steinberg EP, Javitt JC, Sharkey PD, et al: The content and cost of cataract surgery. Arch Ophthalmol 1993;111:1041–1049.

105 Subar AF, Block G: Use of vitamin and mineral supplements: Demographics and amounts of nutrients consumed. The 1987 Health Interview Survey. Am J Epidemiol 1990:132:1091–1101.

106 Taylor A, Jacques PF, Epstein EM: Relations among aging, antioxidant status, and cataract. Am J Clin Nutr 1995;62:1439S–1447S.

107 Taylor A: Nutritional and Environmental Influences on the Eye. Boca Raton, CRC Press, 1999.

108 Taylor A, Hobbs M: Assessment of nutritional influences on risk for cataract. Nutrition 2001;17: 845–857.

109 Taylor A, Jacques PF, Nowell T, et al: Vitamin C in human and guinea pig aqueous, lens and plasma in relation to intake. Curr Eye Res 1997;16:857–864.

110 Taylor HR, West SK: A simple system for the clinical grading of lens opacities. Lens Res 1988; 5:175–181.

111 Taylor HR: Ultraviolet radiation and the eye: An epidemiologic study. Trans Am Ophthalmol Soc 1989;87:802–853.

112 Taylor HR: Epidemiology of age-related cataract. Eye 1999;13:445–448.

113 Taylor HR, Tikellis G, Robman LD, et al: Vitamin E supplementation and macular degeneration: Randomised controlled trial. BMJ 2002;325:11–14.

114 Tessier F, Moreaux V, Birlouez-Aragon I, Junes P, Mondon H. Decrease in vitamin C concentration in human lenses during cataract progression. Int J Vitam Nutr Res 1998;68:309–315.

115 Trevithick JR, Mitton KP: Vitamins C and E in cataract risk reduction. Int Ophthalmol Clin 2000;40:59–69.

116 The Heart Outcomes Prevention Evaluation Study Investigators: Vitamin E supplementation and cardiovascular events in high-risk patients. N Engl J Med 2000;342:154–160.

117 Varma SD, Srivastava VK, Richards RD: Photoperoxidation in lens and cataract formation: Preventive role of superoxide dismutase, catalase and vitamin C. Ophthalmic Res 1982;14: 167–175.

118 Varma SD, Richard RD: Ascorbic acid and the eye lens. Ophthalmic Res 1988;20:164–173.

119 Varma SD, Devamanoharan PS: Oxidative denaturation of lens protein: Prevention by pyruvate. Ophthalmic Res 1995;27:18–22.

120 Wegner A: Kataraktprävention.Therapeutische Ansätze und klinische Betrachtung des Erreichten. Ophthalmologe 2003;100:176–180.

121 West S, Munoz B, Emmett EA, Taylor HR: Cigarette smoking and risk of nuclear cataracts. Arch Ophthalmol 1989;107:1166–1169.

122 West S: Ocular ultraviolet B exposure and lens opacities: A review. J Epidemiol 1999;9:S97–S101.

123 West SK: Daylight, diet, and age-related cataract. Optom Vis Sci 1993;70:869–872.

124 West SK, Valmadrid CT: Epidemiology of risk factors for age-related cataract. Surv Ophthalmol 1995;39:323–334.

125 Wu SY, Leske MC: Antioxidants and cataract formation: A summary review. Int Ophthalmol Clin 2000;40:71–81.

126 Zigler JS Jr, Huang QL, Du XY: Oxidative modification of lens crystallins by H_2O_2 and chelated iron. Free Radic Biol Med 1989;7:499–505.

Carsten H. Meyer, MD
Department of Ophthalmology, Philipps University of Marburg
Robert-Koch-Strasse 4, DE–35037 Marburg (Germany)
Tel. +49 6421 2862616, Fax +49 6421 2865678, E-Mail meyer_eye@yahoo.com

Augustin A (ed): Nutrition and the Eye.
Dev Ophthalmol. Basel, Karger, 2005, vol 38, pp 120–147

....

Nutrition and Retina

Ursula Schmidt-Erfurth

Universitätsklinik der Augenheilkunde und Optometrie, Vienna, Austria

Abstract

The impact of nutrition on manifestation and progression of retinal diseases has become an important, controversial topic within recent years. The awareness of this topic in the general population has increased partially due strong commercial advertisements of supplements and diets. However, many potentially beneficial nutritional effects on retinal diseases have not been proven in prospective clinical trials. It is only for a few relatively rare diseases, such as retinitis pigmentosa or gyrate atrophy, that adjustments in nutrition have been proven effective and widely accepted. However, for the majority of patients with retinal diseases the impact of nutritional factors is still insufficiently understood. Theoretically, supplementation of antioxidants could have a beneficial impact on a wide variety of retinal diseases or as a preventive measure by limiting the degree of oxidative damage. The only prospective, controlled, clinical trial providing proven benefit of antioxidant supplementation for a retinal disease is the Age-Related Eye Disease Study (AREDS). Patients with at least intermediate age-related macular degeneration (AMD) were shown to have a significant benefit with regard to disease progression by supplementing with high-dose antioxidants and zinc. It is however unclear whether other antioxidants, such as lutein or zeaxanthin, may be better and whether a preventive supplementation is useful. Especially studies on patients with diabetic retinopathy have implicated an impact of higher cholesterol levels on the progression of the disease. High-fat diets have been overall associated to a number of retinal diseases. With the current knowledge it seems prudent to advise everyone a balanced, low-fat diet as well as vitamin supplementation within the recommended daily allowance. Smoking is an essential factor for oxidative stress, and its cessation should be recommended to everybody in order to prevent or slow down progression of retinal disease. High-dose antioxidant supplementation according to the AREDS trial should currently only be recommended to non-smokers with at least intermediate AMD. Based on results from experimental studies, further prospective clinical studies are warranted on the prevention and inhibition of disease progression in the most common retinal diseases by nutritional means.

The recent decades have provided significant evidence that nutrition is a key issue to maintain and possibly prolong function and structure of different tissues throughout the human body. An increasing life span and high expectations on quality of life have drawn the general population's interest and awareness toward potential benefits of optimizing nutrition. Nutritional supplementation has gained widespread popularity despite limited evidence-based knowledge.

Malnutrition in general leads to degeneration or to aggravation of preexisting degenerative changes in tissues. An association of nutritional factors to variety of retinal and choroidal diseases has been made. Two of the most common diseases, in which nutrition seems to play an essential role, are age-related macular degeneration (AMD) and diabetic retinopathy/maculopathy. Potential nutritional factors for these diseases are discussed in this chapter.

Age-Related Macular Degeneration

A variety of degenerative changes of the retina are covered by the term 'age-related macular degeneration' (AMD). Degenerative changes occur mostly in the macula, the central and functionally most important retinal area. Characteristic changes include drusen, pigmentation, atrophy and choroidal neovascularization with exudation and bleeding. Some of these changes can however be found in different much less prevalent macular dystrophies, where disease is related to familial tendency or heredity [1, 2].

Retinal degeneration is considered to be secondary to a degree of genetic predisposition and a variety of physical and environmental influences. This chapter will mostly discuss nutrition as an environmental influence in AMD for several reasons. AMD is the most prevalent degenerative retinal disease and is the leading cause for irreversible central vision loss and legal blindness in predominantly Caucasian populations [3, 4]. Progressively older populations in the industrialized nations will lead to even higher prevalence for AMD in the future. Current treatment of AMD is limited to exudative AMD with the aim of preventing further vision loss [5]. Due to the impact of AMD on society, there has been intensified research in the field of AMD. There is some indication that prevention of AMD and prevention of progression of early to intermediate stages of AMD is possible with nutritional supplements.

This chapter will first outline different postulated pathomechanisms for AMD and possible impacts of nutrition on these pathomechanisms. Consequently, postulated preventive mechanisms of vitamins and nutritional supplements in retinal degenerations will be discussed. The third part will discuss the most important and only large prospective clinical trial on prevention of progression of AMD by high-dose nutritional supplementation and the current recommendations for nutritional supplementation in AMD.

Nutrition and Pathomechanisms in AMD

Theories for pathomechanisms in AMD have mostly evolved from histo-logical studies describing characteristic pathologic findings in AMD at differ-ent stages of the disease. In addition, especially recent animal models showing pathologic findings similar to AMD have and will increase the understanding of AMD [6–9]. Epidemiologic studies have provided further information on the incidence and progression of AMD-related lesions.

Currently there are three main theories for retinal changes associated with non-exudative stages of AMD: the oxidative stress theory, the choroidal circu-lation theory and the degeneration of Bruch's membrane theory. None of these theories exclude other theories from being contributory to degenerative changes associated with AMD. There are indications that nutrition and nutritional sup-plements play a role in all of these theories.

Oxidative Stress Theory

The oxidative stress hypothesis for the etiology of AMD is based on the breakdown of protective antioxidant systems within the retina. Halliwell [10] defined antioxidants as any substance that at low concentrations significantly delays or inhibits oxidation. Counterpart to antioxidant systems are free radi-cals, which are molecules or atoms that possess via the process of oxidation at least one unpaired electron. This property makes free radicals highly reactive.

In normal aerobic metabolism, oxygen undergoes typically a four elec-tron reduction to water without production of free radicals. However, within this reductive path, oxygen can also accept an electron from reducing agents and consequently different, very reactive radicals are formed: superoxide, hydrogen peroxide and hydroxyl radicals. The key mechanism by which tissue damage occurs due to reactive radicals is the formation of lipid peroxides. Lipid peroxides are formed when radicals react with unsaturated fatty acids, which are largely present in cells as glyceryl esters in phospholipids and triglycerides. As long as oxygen is present, one free radical can lead to oxida-tion of thousands of fatty acids. A termination reaction, in which two radicals form a non-radical product, can interrupt the chain reaction. Free radical scav-engers can also inhibit further lipid peroxidation. This is the first aspect where nutrition comes into play. Well-known free radical scavengers are vitamin E (α-tocopherol) or vitamin C (ascorbate). They subsequently prevent the for-mation of lipid peroxides.

The second pathway by which lipid peroxidation occurs is photo-oxidation. Light activates oxygen electronically to form singlet oxygen, which also reacts as a radical with unsaturated fatty acids. The formation of non-conjugated hydroper-oxide isomers indicates singlet oxygen-induced damage to unsaturated fatty acids. Carotenoids are known for their capacity to inhibit lipid peroxidation by quenching singlet oxygen. Well-known carotenoids or xanthophylls in the biological system are β-carotene, a precursor of vitamin A, lutein and zeaxanthin.

The retina is susceptible to lipid peroxidation by radicals for different reasons. Polysaturated fatty acids are abundant in the retina and particularly in the macula. They are found in photoreceptor outer segments and are very sensitive to peroxidation in proportion to their number of double bonds [11]. The rod inner segments are also very rich in mitochondria, which might leak activated oxygen due to aerobic metabolism. Photoreceptors degenerate when they are exposed to continuous oxidative challenge or antioxidative defense mechanisms are reduced. The retina and especially the macula are subject to high levels of light exposure. Light and especially wavelength in the blue spectrum are, as described, in the presence of oxygen potent inductors of lipid peroxidation [12]. The excellent oxygen supply through especially the choroidal vascular network elevates the risk for oxidative damage. The oxygen tension is highest at the choroid and decreases towards the photoreceptor inner segments, where metabolic demand for oxygen is highest [12].

The retina has several defense mechanisms against the production of free radicals. The retinal pigment epithelium (RPE) is most important for retinal protection to oxidative stress due to its numerous antioxidant enzymes. These enzymatic systems are however very susceptible to dietary deficiency. Micronutrients such as selenium, zinc, manganese and copper are essential for antioxidant enzymes [13]. Free radical scavenging is another protective mechanism. It involves antioxidant nutrients such as vitamin E (α-tocopherol) [14–17] and vitamin C (ascorbate) [18, 19]. Quenching of singlet oxygen by carotenoids such as β-carotene, lutein and zeaxanthin is an important mechanism to inhibit light-induced oxidative damage. Further defense mechanisms include antioxidant compounds such as metallothionein [20], melanin, glutathione and DNA repair.

The theory that oxidative stress or reduced antioxidant mechanisms, possibly induced by insufficient diet, contributes to AMD is based on several animal and human studies. Animals with deficiency of vitamin E have shown increased retinal damage being exposed to light, while supplementation of vitamin C showed reduced retinal damage [21, 22]. Dietary insufficiency of vitamin C in guinea pigs without any additional oxidative stress leads to an increase in retinal lipid peroxidation [23]. Melatonin produced mainly by the pineal gland is also found in photoreceptor and proves antioxidant properties [24, 25]. There is increasing evidence that β-carotenoids play an essential role in preventing oxidative degeneration. Under all conditions of free radical-initiated autoxidation of carotenoids, the breakdown of β-carotene was much faster than that of lutein and zeaxanthin [26]. The slow degradation of the xanthophylls zeaxanthin and lutein may be suggested to explain the majority of zeaxanthin and lutein in the retina of man and other species. In correspondence to that, the rapid degradation of β-carotene under the influence of natural sunlight and UV light is postulated to be the reason for the almost lack of those two carotenoids in the human retina [26]. In a recent animal model it was shown that supplementation of zeaxanthin reduces photoreceptor degeneration [27, 28]. Measurement of carotenoids in the living primate eye using resonance Raman spectroscopy demonstrated progressive reduction of carotenoids in ageing primates [29].

Recent studies in humans have shown a close correlation of decrease in retinal melatonin, lutein and zeaxanthin to increasing age, the most important cofactor for AMD [30, 31]. The currently strongest evidence for a correlation of oxidative stress and AMD comes from the AREDS. It investigated the use of high-dose vitamin C and vitamin E, β-carotene (lutein and zeaxanthin were not commercially available at the beginning of the study), zinc and copper on AMD and cataract. Patients with intermediate dry AMD or significant vision loss due to AMD in the second eye showed most risk reduction for progression to advanced AMD and for a three line decrease in visual acuity [32].

Choroidal Circulation Theory

Already in 1937, Verhoeff and Grossman [33] linked sclerotic changes in the choriocapillaris to AMD. It was further postulated that removal of waste materials and the supply of nutritional substances as well as oxygen exchange to the neural retina is impaired due to choroidal vascular alterations [34]. In several studies using different techniques it has been shown that reduced choroidal blood flow is associated with AMD [35–38]. This theory is supported by correlation of AMD with diseases involving arteriosclerotic changes such as the carotid artery disease [38]. Interestingly, most epidemiological studies, except for the Rotterdam Study [39], did not find a direct association of cardiovascular disease and AMD [40–42], however many factors for cardiovascular disease are also risk factors for AMD. One of the most important risk factors for AMD, which is consistently found in epidemiologic studies, is smoking. Smoking increases the risk for all types of AMD significantly [40, 43–45], for neovascular AMD even up to 6.6-fold [46]. Hypertension, a common risk factor for many vascular diseases, has been found associated to AMD [42, 45, 47] as well as a higher body mass index (BMI) [41, 48]. High blood lipid levels are a risk factor for arteriosclerosis and AMD [40, 42]. Especially high serum levels of low density lipoprotein (LDL), very low density lipoprotein (VLDL) and triglycerides have been found to be an indicator for an increased risk for AMD [49]. Prospective studies showed that total fat intake was positively associated with risk of AMD [47, 50]. Higher intake of specific types of fat, including vegetable, monounsaturated, and polyunsaturated fats and linoleic acid, rather than total fat intake may be associated with a greater risk for advanced AMD. Diets high in ω–3 fatty acids, fish, nuts and high levels of serum high density lipoproteins (HDL) were inversely associated with risk for AMD [47, 50, 51]. If high levels of serum LDL, VLDL and triglycerides are associated with AMD, lipid-lowering medications may have a beneficial effect as shown by McCarty et al. [44]. Consumption of coffee and caffeine could potentially play a role in this model. Caffeine has a vasoconstrictive effect and could potentially alter choroidal blood flow [52]. Epidemiologic evaluation within the Beaver Dam Eye Study did however not demonstrate a correlation of coffee or caffeine intake to AMD [53]. Hyperhomocysteinemia is another independent risk factor strongly associated with cardiovascular disease [54, 55]. Hyperhomocysteinemia is associated with alterations in vascular morphology, loss of endothelial antithrombotic function, and induction of a procoagulant environment. A recent clinical study revealed and association of exudative AMD with higher levels of homocysteine [56].

Many of these proarteriosclerotic influences also have an effect on the RPE. It has been suggested and demonstrated in experimental work [57] that choroidal vascular changes can occur secondary to RPE atrophy. Possibly changes in Bruch's membrane lead to reduced signaling between RPE and choriocapillaris and to regression of the choriocapillaris. However, nutrients which influence vascular changes throughout the vascular system may have an impact on the development of AMD, regardless of vascular changes in the choriocapillaris being a primary or secondary event.

Bruch's Membrane Theory

Age-dependent changes in Bruch's membrane are considered to compromise transport of nutrients and metabolic substances from and to the RPE. Degeneration and thickening of Bruch's membrane initiate or at least contribute to AMD. The site of main resistance is the inner collagenous layer. With increasing age, hydrophilic properties of Bruch's membrane decrease exponentially. This is explained by increased lipid deposits in Bruch's membrane [58]. Interestingly there is with age a significant increase of peroxidized lipids, which are mostly derived from long-chain polyunsaturated fatty acids, particularly docosahexaenoic acid and linolenic acid [59]. These fatty acids are commonly found in photoreceptor outer segments, but also higher nutritional consumption of linoleic acid was associated with a higher risk for AMD [51]. Drusen are the characteristic changes of early AMD, being round, well-defined or confluent, yellowish deposits of varying size on ophthalmoscopy. These deposits exceed those related to the normal ageing process in size and are located between the inner collagenous layer of Bruch's membrane and the basement membrane of the RPE. The origin of drusen components could basically be the RPE or the choroid. All waste products from the RPE metabolism are transported via Bruch's membrane to the choroidal circulation, which supports the hypothesis of RPE origin. Furthermore, studies of drusen components have shown that major components of drusen are fatty acids and phospholipids rather than cholesterol, indicating the RPE rather than the choroid as origin. Despite the fact that degenerative changes of the RPE are regarded the primary cause for the formation of drusen [60, 61], there is indication that a variable degree of drusen composition are of choroidal origin. Curcio et al. [62] have drawn a strong correlation of some cholesterol components of drusen to arteriosclerosis, and animal models have shown that nutritional cholesterol can accumulate in Bruch's membrane [63]. Interestingly, apolipoprotein E (apoE)-deficient mice and mice on a high-fat diet show deposits in Bruch's membrane that are similar to AMD [8, 64]. ApoE is considered an important regulator of lipid trafficking in AMD and was recently shown to be effected by HDL [65]. The apoE gene polymorphism was found in a genetic study to have a significant association with the risk for AMD [66]. Overall, changes in Bruch's membrane seen with age and in AMD seem closely related to degenerative changes in the RPE, and its ability to degrade especially peroxidized lipids is possibly influenced by proarteriosclerotic conditions. It is possible that limitation of peroxidation by antioxidants and reduction of proarteriosclerotic nutritional factors could have a beneficial effect on changes in Bruch's membrane.

Dietary Components with a Possible Role in Retinal Degeneration

Dietary components affecting retinal degeneration, especially AMD, are discussed in this section. These components can be basically divided into beneficial and risk factors. This section elucidates experimental and clinical aspects of different dietary components on ocular health.

Antioxidants
The potential of antioxidants is to scavenger or quench free radicals of different origin. The antioxidants discussed in this chapter are vitamin E, vitamin C and the carotenoids β-carotene, lutein and zeaxanthin. Antioxidants vitamin E and carotenoids predominate in lipophilic cell segments especially in the RPE. Vitamin C, a common antioxidant in the retina is however found, compared to other antioxidants, less commonly in the RPE cells.

Vitamin E
In 1922, vitamin E was first discovered as an essential micronutrient for the reproduction in female rats [67]. Eight different forms of vitamin E are only

produced by plants. Interestingly, there are specific human binding proteins especially for α-tocopherol, which seems to be the preferentially used form of vitamin E by the human body. The RPE and the outer photoreceptor segments of rods have the highest concentration of vitamin E [68], which can be increased by dietary supplementation as shown in animals [68]. Increased concentrations of vitamin E in photoreceptors and RPE were also found after increased oxidative stress to ocular tissues by e.g. constant light exposure [69]. This indicates that the retina is capable of upregulating vitamin E for protection from oxidative stress and may also be the reason for inconsistent data on retinal damage to oxidative stress. Dietary deficiency of vitamin E has led to inconsistent results in animal models, but several papers report on lipofuscin accumulation in the RPE and photoreceptor loss. These findings are comparable to retinal changes seen in AMD.

There is a strong association of vitamin E and retinal degeneration found in untreated patients with inherited abetalipoproteinemia. In this condition fat-soluble vitamins cannot be absorbed sufficiently and characteristic retinal changes including lipofuscin deposits occur [70]. Interestingly, only the supplementation of vitamin E prevented retinal degeneration in this condition [71]. In addition to its antioxidant activities especially α-tocopherol has a protein kinase C (PKC)-inhibiting effect [72] which is related to an antiatherosclerotic effect. The possible correlation of AMD and arteriosclerosis has been elucidated previously.

The majority of epidemiologic studies have found a correlation of vitamin E and AMD [73–76]. A vitamin E-rich diet and higher serum levels of vitamin E seem to be related to a lower risk of AMD, however results were mostly not statistically significant. Epidemiologic studies on vitamin E have been difficult to evaluate because there are several potentially influencing factors. Dietary and supplemented vitamin E has been shown to be different with regard to the subforms. In addition, even subjects supplemented equally with α-tocopherol have shown significant variations in α-tocopherol plasma concentrations [77].

There have been several prospective randomized trials to evaluate the role of vitamin E supplementation in AMD [32, 78–80], however in most of these trials, vitamin E was given with other antioxidants and nutritional supplements. The Vitamin E, Cataract and Age-Related Maculopathy Trial (VECAT) is currently the only trial reporting on supplementation of daily 335 mg α-tocopherol alone [78]. 1,193 participants were randomized into two groups and followed up for 4 years. Two other trials have investigated lower doses of vitamin E (daily dose of 40–50 mg in addition to β-carotene ± vitamin C) [79, 80]. None of these three trials showed any significant benefit of antioxidant therapy on progression and prevalence of AMD, however none of these trials was likely to be sufficiently powered. The only trial that demonstrated a statistically significant benefit for antioxidant supplementation containing vitamin E for certain subgroups was the AREDS [32].

Only sporadic side effects have been noted with higher doses of vitamin E. However, vitamin E has an antiplatelet and anticoagulant effect and aggravates vitamin K deficiency, resulting in inhibition of coagulation. The suggested safe upper limit of *d*-α tocopherol is 540 mg/day.

Vitamin C

Vitamin C (*L*-ascorbic acid) is a water-soluble antioxidant reacting directly with hydroxyl radicals, superoxide and singlet oxygen. It is found throughout the retina, but less commonly in the RPE. Evidence that vitamin C may protect against retinal light damage is derived from a series of animal studies. Supplementation of vitamin C led to an increased level of vitamin C in the retina and to reduced light-induced retinal damage [21, 81]. Epidemiologic studies reported on an increased risk for AMD with low plasma levels of vitamin C, however high plasma levels were not found to be protective [73–76, 82, 83]. Two randomized controlled trials tested nutritional supplementation of vitamin C in addition to other supplements. The very small Visaline trial did not show inhibition of AMD progression for patients receiving vitamin supplementation [80]. The AREDS trial supplementation regimen contained 113 mg of vitamin C. The AREDS regimen with additional vitamin E, β-carotene, zinc and copper showed a significant benefit with regard to disease progression in certain subgroups. Antioxidant supplementation alone was however not as effective as in combination with zinc [32].

Common are gastrointestinal adverse reactions at levels in excess of 1,000 mg/day. Patients with hemochromatosis, thalassemia and urinary/renal stones are at particular risk consuming >1,000 mg vitamin C per day.

Carotenoids

The three major carotenoids discussed in this section are lutein, zeaxanthin and β-carotene. Lutein and zeaxanthin are the two carotenoids that concentrate most in the retina. Lutein spreads more diffusely throughout the macula, zeaxanthin, a stereoisomer of lutein, accumulates in the foveal center. β-Carotene, a precursor of vitamin A, is usually not found in higher concentrations in the retina, but high-dose oral supplementation can increase retinal β-carotene levels [32]. There is increasing evidence that antioxidant properties of carotenoids have a protective role in the retina.

β-Carotene

β-Carotene, a hydrocarbon carotenoid, is one of the major carotenoid precursors of vitamin A. In contrast to vitamin A, which has no capacity to quench singlet oxygen and only limited free radical scavenging effect [84], β-carotene has been proven to be an effective antioxidant [85]. However, its antioxidant capacities are considered inferior to the xanthophylls lutein and zeaxanthin. It is interesting that β-carotene is mostly found in the RPE/choroid and not in the overlying retina [86]. This finding is suggestive for metabolic cleavage of this carotenoid to vitamin A in the RPE/choroid complex. β-Carotene is mostly found in marigold flowers, dark green vegetables and colored fruit [87]. Dietary hydrocarbon carotenoids, such as β-carotene, have been studied in different populations [40, 76, 83] and were shown to be protective. Provitamin A carotenoids were associated with a 2-fold lower incidence of large drusen over 5 years [76]. Similar findings were reported in the Eye Disease Case-Control Study [83]. However, this association was less evident when correcting for possible influences of other dietary constituents. Evaluation of serum levels of carotenoids eliminates estimations on dietary β-carotene and variability of absorption. However, results in epidemiologic studies evaluating serum levels of carotenoids have been variable and studies are often not comparable. There seems however to be a trend that low serum levels of carotenoids are related to a higher risk of AMD [74] and higher serum levels to a lower risk of different stages of AMD [73, 82].

There have been three randomized controlled trials substituting oral β-carotene of 15–40 mg in addition to other supplements. The two studies showing no significant difference in prevalence and progression of AMD were statistically not sufficiently powered [79, 80]. The AREDS regimen, which included 15 mg of β-carotene, showed however a significant benefit, limiting the progression of AMD for certain subgroups [32].

The most important risk of supplementing β-carotene has been reported in two studies relating β-carotene supplementation (20–30 mg/day) in smokers to a higher risk for lung cancer [88, 89]. Carotenoids are generally non-toxic, because the efficiency of absorption decreases rapidly as supplementation dose increases. In addition, conversion to vitamin A is not rapid enough to cause toxicity. Hypercarotenemia (>30 mg/day) is characterized by yellow skin colorations.

Lutein and Zeaxanthin

In 1945 it was first reported that the macula lutea could, based on spectrum analysis, contain carotenoids [90]. Bone et al. [91] confirmed the findings and demonstrated that the xanthophylls lutein and zeaxanthin were both present in the macula. Zeaxanthin predominates in the central fovea and lutein more in the periphery [92]. The two major structures where lutein and zeaxanthin are found are the Henle fiber layer and RPE. This correlates with the two assumed major functions of lutein and zeaxanthin. The high absorptivity of these pigments in the inner retina functions as an efficient filter for blue light. Lutein has an absorption maximum of about 445 nm and zeaxanthin of about 451 nm. Recently a specific lutein-binding protein was isolated from human retina [93]. Upon binding this protein, lutein exhibits a shift in its absorption maximum to 460 nm, the same wavelength maximum that is determined psychophysically for the macular pigment. It is estimated that macular pigments within the Henle fiber, overlying the photoreceptors, reduce the blue light intensity normally by 40% and can be increased up to 90% [94]. This could significantly reduce oxidative stress on the retina and may be sufficient to explain risk reduction for AMD as shown in epidemiologic studies.

Carotenoids, including lutein and zeaxanthin, are in addition to their blue light-absorbing characteristics known for their antioxidant function [95]. It is hypothesized that carotenoids prevent oxidation of nucleic acids, proteins and polyunsaturated fatty acids by preferential oxidation. The oxidized carotenoid potentially reacts with ascorbate regenerating the unaltered carotenoid [96]. In vivo this mechanism is still controversial, however lutein and zeaxanthin have been identified in photoreceptor outer segments and RPE [86, 97]. This is essential for the presumed antioxidative function of carotenoids because in photoreceptor outer segments and the RPE effects of oxidation are considered to have the most damaging effect. In the outer retina, oxygen partial pressure is very high resulting in a high rate of blue light-induced singlet oxygen formation. It is assumed that lutein and zeaxanthin have a passive antioxidant function by absorbing blue light and an active antioxidant function by preferential oxidation. Potentially, lutein has an additional mechanism to reduce the risk of AMD. Plasma lutein levels have been reported to be directly related to a decline in the progression of intima-mediated thickness in the common carotid artery [98]. Atherosclerosis, as discussed previously, may play role in the pathogenesis of AMD.

Several epidemiological studies indicate a correlation of lutein and zeaxanthin and AMD. The Eye Disease Case-Control Study was first to report an increased risk of neovascular AMD in association with decreased serum levels of carotenoids as well as decreased risk for AMD in correlation with high plasma levels of lutein and zeaxanthin [40, 82]. Intake of dark green, leafy vegetables, a major source for lutein [99], was shown to reduce the relative risk of developing AMD significantly [83]. In contrast, the Beaver Dam Eye Study did not initially show a relationship between carotenoid ingestion and early and late AMD [75], however further reports from this study documented a correlation of provitamin A carotenoids and the incidence of large drusen [76]. Results from donor eye studies indicate a significantly decreased risk for AMD in case of high macular levels of lutein and zeaxanthin. Newer imaging techniques such as resonance Raman spectroscopy and autofluorescence are providing further evidence that macular pigment density measurements can be related to AMD [30, 100] and that increased oral supplementation of lutein can increase macular pigment density [101].

However, there is still no direct evidence that dietary lutein and zeaxanthin in the form of green, leafy vegetables or corn or lutein supplementation can reduce the risk for AMD. Currently there are no results from a prospective randomized study. Unfortunately, the AREDS was not able to integrate lutein and zeaxanthin in their study due to unavailability of both carotenoids in a commercial formulation. Subsequently, β-carotene was used even though this carotenoid is typically not found in either the lens or the retina.

Potential risks of high-dose oral intake/supplementation of lutein and zeaxanthin are difficult to judge. So far, no risks on high-dose supplementation of these carotenoids have been reported, even in South Pacific islanders consuming as much as 27 mg/day. However, the carotenoid β-carotene was linked to an increased risk of lung cancer in smokers. Therefore, potential risks of high-dose supplementation of lutein and zeaxanthin should be ruled out before recommending such therapy. Levels of up to 6 mg of lutein and zeaxanthin per day, as found in the upper quintile of the Eye Disease Case-Control Study Group, seem to be, given the current sate of knowledge, reasonable dietary target levels. Preferentially these levels should be obtained via dietary sources such as spinach, kale, broccoli, peas and brussels sprouts for lutein and corn, orange peppers, oranges and honeydew for zeaxanthin.

Further prospective randomized studies will be necessary before high-dose lutein and zeaxanthin supplementation should be recommended to reduce the risk of degenerative retinal diseases.

Cofactors of Antioxidant Enzymes

Nutrients as essential minerals (i.e. zinc, copper, iron, selenium, etc.) may play an important role by their involvement in the enzymatic antioxidant system. Many enzymes such as superoxide dismutase, catalase, glutathione peroxidase, are essential to quench free radicals and other reactive oxygen species (ROS). However, many enzymes require essential minerals as cofactors. A deficiency of these nutritional cofactors negatively affects the antioxidant capacity of these enzymatic systems. The upregulation of antioxidant enzyme systems has been related to AMD in humans [102]. Certain baseline levels of cofactors are necessary to permit enzyme activity, however higher levels of cofactors do not improve enzymatic activity. Therefore an increase of cofactor levels significantly above baseline levels by nutritional supplementation does not seem to improve oxidative protection.

Zinc is the most extensively studied enzymatic cofactor in the retina and the RPE. Its retinal concentration is higher than in all other body parts [103]. It acts as a cofactor for the retinal dehydrogenase and catalase [104]. Zinc was however also shown to be important in other non-antioxidant enzymatic systems. A severe deficiency of zinc in rats showed significant degenerative changes in RPE and photoreceptor outer segments [105]. Low levels of zinc in parenteral nutrition in humans led to reversible changes in the electroretinogram [106]. The epidemiologic Beaver Dam Eye Study evaluated zinc intake and macular pigmentary changes and found by comparing the highest and lowest quintile of zinc intake less prevalent and less newly developed pigmentary abnormalities in the higher quintile [74, 76]. There have been three prospective controlled studies on supplementation of zinc. Two studies supplementing 200 mg zinc per day demonstrated controversial results on progression of AMD [107, 108]. The AREDS included about 80 mg of zinc per day in two of their regimens. One of the four arms of the study supplemented zinc only. In this group the probability of a defined AMD or visual acuity event at 5 years' follow-up was decreased by 6.2 and 3.6% compared to placebo. However, combination with antioxidants further improved the outcome [32].

Copper has been found to concentrate also in ocular structures and is a coenzyme for superoxide dismutase, an antioxidant enzyme, which has been found to increase with age in AMD models [109]. In addition, copper is considered to be involved in the metabolism of the RPE [110]. However, studies in humans evaluating copper and AMD have been limited to higher ceruloplasmin levels in AMD patients [111]. Copper has been included into the AREDS treatment arms receiving zinc, not for a potential benefit preventing progression of AMD, but to prevent possible zinc-induced copper deficiency anemia.

Other essential minerals are considered to be important cofactors for enzymatic antioxidant systems, however hints for an association to retinal degenerations are limited.

Folic Acid

Even though folic acid is not directly associated with retinal degeneration, it may be an important nutritional aspect in the context of AMD. Inadequate intake of folic acid and, to a lesser extent, vitamin B_6 and vitamin B_{12}, increases homocysteine levels and substitution of folic acid reduced homocysteine levels [112]. Several studies have shown that hyperhomocysteinemia is an independent risk factor for atherosclerosis, coronary heart disease and venous thromboembolism [113, 114]. The association between AMD and atherosclerosis remains controversial, but hyperhomocysteinemia was shown to be not only a risk factor for retinal vascular occlusive disease [115], but a recent study also indicated an association to neovascular AMD. Hyperhomocysteinemia $>15 \mu mol/l$ was found in 44.7% of patients with neovascular AMD. These results were statistically significant compared to dry AMD and unaffected control patients [56]. Currently there are no prospective controlled clinical studies on the impact of folic acid on incidence and progression of AMD. The currently upper limit recommended for folic acid is 1 mg/day.

Nutritional Risk Factors

Dietary Fat Intake
Epidemiologic studies indicate a possible role of cardiovascular risk factors in patients with AMD, suggesting that the two diseases may be interrelated [116]. There are three basic mechanisms by which dietary fat intake could increase the risk for AMD. Dietary fat has been clearly associated with atherosclerosis and could therefore affect blood supply in the choroid and retina. Higher blood levels of fatty acids might also increase oxidative damage due to high susceptibility of fatty acids to oxidation especially under high oxygen tension and light exposure as found in the macula. Alternatively, increased deposition of fat in Bruch's membrane has been postulated [62, 63], that would adversely affect exchange of nutrients and waste products to and from the RPE. The Beaver Dam Eye Study evaluated several aspects of dietary fat intake. People with intakes of saturated fat and cholesterol in the highest quintile had a greater risk of early age-related maculopathy [117]. Special interest has been on serum cholesterol levels. The Eye Disease Case-Control Study showed people with an increased serum cholesterol level to be compared to people with low levels of serum cholesterol at higher risk for neovascular AMD [40]. Despite few older studies showing no correlation or inverse correlation of serum cholesterol levels and AMD [118, 119], there is growing evidence for a correlation of dietary fat intake and serum cholesterol levels to AMD [50, 51, 120]. However, special focus should be given to types of dietary fats. Saturated, monounsaturated, polyunsaturated, transunsaturated fats and linolenic acid were found to increase the risk of progression or development of AMD [47, 50, 51]. Especially transunsaturated fatty acids were shown to affect blood lipid levels by increasing LDL and decreasing HDL [121]. Several studies indicated that statins (hydroxymethyl glutaryl coenzyme A reductase inhibitors), used to lower LDL serum cholesterol levels to prevent cardiovascular events, could reduce the risk of especially neovascular AMD [122, 123]. In the Beaver Dam Eye Study however no association of statins and any stage of AMD was found [124].

Certain foods associated with dietary fat intake have however shown a beneficial effect in AMD and cardiovascular disease. The consumption of nuts was associated with a reduction of risk of AMD progression [47] and was found to be protective in cardiovascular disease [125]. The beneficial effect can be explained by the bioactive compound resveratrol, which has antioxidant, antithrombotic and anti-inflammatory properties [126]. Fish intake has also been related to a decreased risk for progression of AMD [47]. Long chain ω–3 fatty acids, especially docosahexanoic acid found primarily in fish, are presumed to be the beneficial component.

It is interesting that studies have shown obesity, also a risk factor for cardiovascular disease, to be a risk factor for AMD independent of dietary fat intake [47, 48] and that physical exercise can reduce the risk.

Smoking

Even though smoking does not directly fit into the category of nutritional risk factors, it is one of the most important risk factors known for AMD. Tobacco smoke has been proven to contain numerous toxic and cancer-inducing agents and has shown to be a major risk factor for cardiovascular disease. The highly toxic nicotine alone could explain why smoking is a major risk factor for AMD. Through its vasoconstrictive effect it may have direct effects on the choroidal circulation. In addition, nicotine reduces serum levels of antioxidants. Furthermore, in animal models, smoking led to sub-RPE deposits.

The majority of epidemiological studies found a statistically significant relationship between smoking and development or progression of AMD. Most of these studies found an odds ratio (OR) of 2:3 [46, 127–130]. The risk seems to be even higher for progression to neovascular AMD. Even though a very slowly decreasing risk for AMD was seen after smoking was stopped, it was found to be still significantly increased compared to people who never smoked [46, 127]. Even up to 20 years after cessation of smoking an increased risk for late AMD was reported in some studies [127, 130]. A dose relation between smoking and AMD has been suggested, especially in persons with neovascular AMD [46, 127, 129].

It is out of question that cessation to stop smoking is recommendable for everyone, but especially patients with earlier stages of AMD should be made aware of the high risk to progress to neovascular AMD. In addition, high-dose vitamin supplementation containing β-carotene, as found effective for certain stages of AMD in the AREDS trial, is not recommended for any smoker due to an increased risk for lung cancer [88, 89].

Alcohol

For cardiovascular disease it has been widely accepted that moderate consumption of alcohol has a protective role [131, 132]. The unclear issues however still remain, including the role of beverage type, pattern of drinking, and the risk that moderate drinking can lead to problem drinking.

The first National Health Nutrition and Examination Survey (NHANES-1) found that moderate wine consumption is associated with decreased odds of developing AMD [133]. The negative association of moderate alcohol consumption and AMD was found to have a statistically significant OR of 0.86. Persons drinking wine seemed to have the highest risk reduction [133]. Other studies did not support these findings [134, 135].

Alcohol by itself has favorable effects on the level of HDL cholesterol and inhibition of platelet aggregation. Wine, particularly red wine, has high levels of phenolic compounds that favorably influence multiple biochemical systems, such as increased HDL cholesterol, antioxidant activity, decreased platelet aggregation and endothelial adhesion [136]. One of these phenolic compounds in red wine, which has been studied extensively and has been mentioned already earlier, is resveratrol.

Evidence for a protective effect of alcohol in AMD is weak, however there is some indication that moderate consumption, especially of red wine, possibly also due to additional phenolic compounds, could have some protective effect in AMD.

The Age-Related Eye Disease Study

Several aspects led the National Eye Institute to initiate a clinical trial on high-dose supplementation with vitamin C and vitamin E, β-carotene, and zinc for AMD (fig. 1). AMD had become the leading cause of visual impairment and blindness in many industrialized countries among people 65 years or older [137–139]. As discussed, there was inconsistent evidence on the effect of high-dose antioxidants and zinc, and increasing public demand for unproven supplementation led to a public health concern. The multicenter, prospective, randomized study enrolled 3,640 patients aged 55–80 years with AMD from 1992 to 1998 [32]. The average follow-up was 6.3 years. Four major AMD eligibility categories were established (table 1). Only participants having at least extensive small drusen ($\geq$ category 2) were randomly assigned to one of four treatment arms: (1) vitamin C (500 mg/day) + vitamin E (400 IU/day) + β-carotene (15 mg/day); (2) zinc oxide (80 mg/day) + copper (2 mg/day); (3) regimens 1 + 2 (table 2), and (4) placebo.

Cupric oxide was added to the tablet formulation to offset the risk of copper deficiency anemia. Vitamin C, vitamin E and zinc were included at much higher levels as the recommended daily allowance (RDA) published by the Food and Nutrition Information Center [see www.nal.usda.gov]. Currently there is no RDA for β-carotene. Standardized additional multivitamin and mineral supplement at no more than RDA levels was supplied to 95% of participants.

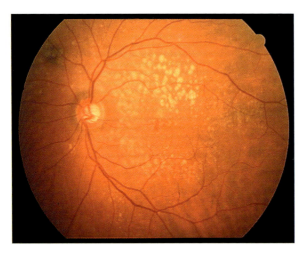

Fig. 1. Intermediate age-related macular degeneration showing large soft drusen. The AREDS trial showed a benefit of high-dose antioxidants plus zinc for non-smoking patients with this stage of disease.

Table 1. AREDS AMD eligibility categories

Category	Definition
1	No drusen or drusen <63 μm with an area <125 μm diameter circle and no pigment abnormalities
2	Small drusen (<63 μm) with an area ≥125 μm diameter circle with possible pigment abnormalities but no geographic atrophy or at least one intermediate size druse (≥63 μm, <125 μm) or no drusen required if pigment abnormalities present
3a	Intermediate size druse (≥63 μm, <125 μm) ≥360 μm diameter circle if soft indistinct drusen are present, ≥656 μm diameter circle if soft indistinct drusen are absent. Pigment abnormalities can be present but no central geographic atrophy or at least one large druse (≥125 μm or no drusen required if non-central geographic atrophy is present
3b	First eye same as category 3a; VA <20/32 in second eye not due to AMD
4a	First eye category 1, 2 or 3a; second eye with advanced AMD
4b	First eye category 1, 2 or 3a; VA <20/32 in second eye due to AMD, but no advanced AMD

Primary outcomes were progression to advanced AMD and at least moderate visual acuity loss from baseline (≥15 letters). Comparison with placebo demonstrated a statistically significant odds reduction for the development of advanced AMD with antioxidants plus zinc (regimen 3) (OR 0.72; 99%

Supplementation	Total daily dose
Vitamin C	50 mg
Vitamin E	400 IU
β-Carotene	15 mg
Zinc	80 mg
Copper	2 mg

Table 2. AREDS formula: AREDS high-dose supplementation

confidence interval (CI) 0.52–0.98). Supplementation with antioxidants or zinc alone did not demonstrate a significant odds reduction. However, it was found that participants with extensive small drusen, non-extensive intermediate drusen or pigment abnormalities (category 2) had only a 1.3% 5-year probability to progress to advanced AMD. By excluding category 2, patients' significant odds reduction was seen also for antioxidants or zinc alone. A significant odds reduction for at least moderate vision loss was only seen in patients assigned to antioxidant and zinc (regimen 3) (OR 0.73; 99% CI 0.54–0.99). The study was able to show an increase of serum levels for all supplements given in each treatment arm ranging from 18% for zinc to 485% for β-carotene at 1 year. Serum levels at 5 years were slightly lower.

No significant adverse experiences were reported and no significantly decreased or increased risk of mortality was found. However, analysis of zinc vs. no zinc suggested a benefit in mortality risk reduction. The small number of participants dying from lung cancer showed no statistically significant difference by treatment.

This study gives indications for high-dose supplementation of antioxidants and zinc in patients with AMD. However, there are some downsides to the study. It remains unclear to what extent vitamin C, vitamin E and β-carotene are beneficial. The more promising carotenes lutein and zeaxanthin were not evaluated due to commercial unavailability at initiation of the study. AMD categories used in the study (table 1) are not very useful in clinical routine. But furthermore, it remains unclear if progression to AMD can be avoided in the general population by high-dose supplementation of antioxidants and zinc.

Conclusion

Unfortunately there is overall only very limited evidence-based data on the impact of nutritional factors on retinal function and prevention of degenerative changes in the retina. There is a wide variety of theories emphasizing prevention of oxidative damage. Especially antioxidants are considered potentially beneficial. Some of these antioxidants (lutein, zeaxanthin) may have further blue light-absorbing capacities, which could prevent possible retinal light damage. In addition to the theories emphasizing retinal oxidative damage as a major component for retinal degeneration, there is increasing evidence of correlations, especially with regard to risk factors, relating atherosclerosis to AMD. Association of atherosclerosis nutritional risk factors has been well established.

It is very difficult do give recommendations for nutritional prevention of retinal degeneration. Further controlled studies with long-term follow-up are necessary to give recommendations on supplementation of nutrients above RDA levels for the general population to prevent retinal degeneration. However, a healthy diet, rich in green, leafy vegetables, nuts and fish, avoidance of high-fat diets especially rich in saturated, monounsaturated, polyunsaturated, transunsaturated fats and linolenic acid, and regular consumption of a multivitamin and mineral supplement within the RDA limits could potentially prevent or delay retinal degenerations. It is further strongly recommended to stop smoking, especially when there is a family history of AMD. There is still controversy on recommending moderate alcohol consumption. Red wine in particular seems to have a preventive effect not only on AMD but also on cardiovascular disease. Supplementation of higher doses of folic acid to reduce homocysteinemia, a probably important risk factor in a variety of diseases including AMD, is still controversial, however potential risks of higher dose supplementation of folic acid seem to be very limited.

High-dose supplementation of antioxidants and cofactors is currently not recommended for the general population. No smoker should be on a high-dose β-carotene supplementation as there is strong evidence for an increased risk of lung cancer. However, non-smokers having at least moderate AMD (AREDS grade 2) have been shown to significantly benefit from a high-dose supplementation of vitamin C and vitamin E, β-carotene, zinc and copper in the AREDS trial. Only the studied dosages and supplements should be recommended for AMD patients.

Further controlled studies, especially on lutein, zeaxanthin and folic acid, will hopefully give additional information on preventive nutritional supplements. It is however very questionable if any future study will be large enough to give recommendations on high-dose supplementation for prevention of AMD in the general population.

Diabetic Retinopathy/Maculopathy

Prospective, controlled clinical studies have confirmed that the most effective treatment for diabetic retinopathy (fig. 2) is a tight glycemic control

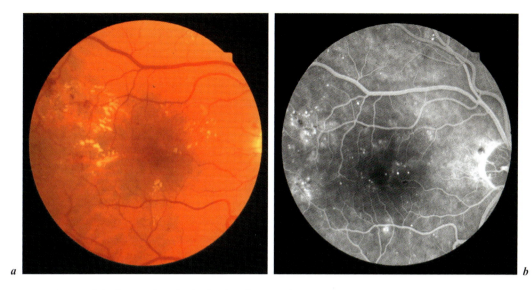

a   b

Fig. 2. Diabetic maculopathy in fundus photography and early angiographic frames. Experimental research demonstrates antioxidants to be effective in inhibiting VEGF-associated diabetic retinal changes. Hard exudates correlate to serum cholesterol levels in type 1 diabetic patients.

[140, 141]. The Diabetes Control and Complication Trial (DCCT) documented the adjusted mean risk for developing any retinopathy to be reduced in comparison to a conventionally treated control group by 76% (p ≤ 0.002). Patients assigned to intensive glucose control in the United Kingdom Prospective Diabetes Study (UKPDS) had a 25% risk reduction in microvascular complications (p = 0.0099). Adjusting nutrition in addition to intensified medical management is essential to obtain a tight glycemic control, however this aspect will not be part of the chapter's discussion.

The two essential aspects discussed in the context of nutrition and diabetic retinopathy/maculopathy are lipid control and antioxidants.

High cholesterol levels have been associated with an increased risk for cardiovascular morbidity. In the Wisconsin Epidemiologic Study of Diabetic Retinopathy (WESDR) there was a significant trend for increasing severity of diabetic retinopathy and of retinal hard exudates with increasing serum cholesterol levels in insulin-using persons [142]. Patients in the Early Treatment of Diabetic Retinopathy Study (ETDRS) with elevated serum cholesterol or LDL levels at baseline were twice as likely to have hard exudates and more likely to develop hard exudates during the study compared to patients with normal serum levels. The risk of losing visual acuity was further associated with the extent of hard exudates even after adjusting for the extent of macular edema [143]. There is indication that for patients with type 1 diabetes severity of retinopathy is positively associated with serum triglycerides and negatively associated with serum HDL cholesterol [144]. First clinical trials using HMG-CoA reductase inhibitors to lower cholesterol levels showed reduction of the progression of retinopathy in diabetic patients with hypercholesterolemia [145]. Further, larger, prospective studies are necessary to evaluate the effects on diabetic retinopathy of lipid-lowering drugs. However, a low-fat and low-cholesterol diet seems to be an overall beneficial approach for patients with diabetic retinopathy.

There is increasing evidence indicating a role for advanced glycation end-products (AGEs) in the development of diabetic retinopathy by inducing blood barrier dysfunction. A direct stimulatory effect of AGEs on expression of retinal vascular endothelial growth factor (VEGF) has been documented [146]. ROS are important in VEGF signaling in vascular cells and the induction of intercellular cell adhesion molecule-1 (ICAM-1) by VEGF involves stress-sensitive pathways [147]. ICAM-1 has been considered a key factor of leukostasis and monocyte adhesion and consequently increased vascular permeability. Experimental studies indicate that antioxidants such as vitamin E and α-lipoic and taurine attenuate upregulation of VEGF and consequently prevent enhanced leukostasis [148, 149]. One of the consequences of diabetes-associated increased serum glucose levels is oxidative stress resulting from increased production of ROS and insufficient upregulation or downregulation of antioxidative defense mechanisms [150]. A strong correlation of lipid peroxidation products, indicating oxidative stress, and VEGF concentrations in the vitreous of patients with proliferative diabetic retinopathy highlight the clinical context [151].

Long-term administration of antioxidants to diabetic rats inhibited in comparison to control animals the development of acellular capillaries significantly and increasing the diversity of antioxidants seemed to provide more protection [152]. Vitamin E and selenium or taurine supplementation further demonstrated reduced biochemical retinal alterations in diabetic rats with poor metabolic control [153].

Significant evidence of protective effects of antioxidants in diabetic retinopathy, as implicated by experimental studies, was however not found in epidemiologic studies [154–156]. Clinical, prospective study results are very limited. A small prospective, placebo-controlled study demonstrated a significant reverse of abnormal retinal blood flow in diabetic patients with minimal diabetic retinopathy being treated with high-dose vitamin E (1,800 IU/day) for 4 months [157].

Further large, prospective, clinical studies are required to determine the effectiveness of high-dose antioxidant supplementation.

Conclusion

Currently a tight glycemic control by nutritional and medical means is most advisable in patients with diabetic retinopathy. Low cholesterol and triglyceride serum levels seem to be beneficial, but the value of statins has to be further evaluated. A low-fat diet is however recommended. From an ophthalmological point of view, high-dose antioxidant nutritional supplementation seems to be promising, but is currently not recommended due to lack of prospective clinical studies.

References

1   Allikmets R, et al: Mutation of the Stargardt disease gene (ABCR) in age-related macular degeneration. Science 1997;277:1805–1807.
2   Heon E, et al: Linkage of autosomal dominant radial drusen (malattia leventinese) to chromosome 2p16-21. Arch Ophthalmol 1996;114:193–198.
3   Klein R, et al: The five-year incidence and progression of age-related maculopathy: The Beaver Dam Eye Study. Ophthalmology 1997;104:7–21.
4   Vingerling JR, et al: The prevalence of age-related maculopathy in the Rotterdam Study. Ophthalmology 1995;102:205–210.
5   Verteporfin in Photodynamic Therapy Report 2: Verteporfin therapy of subfoveal choroidal neovascularization in age-related macular degeneration: Two-year results of a randomized clinical trial including lesions with occult with no classic choroidal neovascularization. Am J Ophthalmol 2001;31:541–560.
6   Ambati J, et al: An animal model of age-related macular degeneration in senescent Ccl-2- or Ccr-2-deficient mice. Nat Med 2003;9:1390–1397.
7   Cousins SW, et al: The role of aging, high fat diet and blue light exposure in an experimental mouse model for basal laminar deposit formation. Exp Eye Res 2002;75:543–553.
8   Dithmar S, et al: Ultrastructural changes in Bruch's membrane of apolipoprotein E-deficient mice. Invest Ophthalmol Vis Sci 2000;41:2035–2042.
9   Weber BH, et al: A mouse model for Sorsby fundus dystrophy. Invest Ophthalmol Vis Sci 2002;43:2732–2740.
10 Halliwell B: Antioxidant defence mechanisms: From the beginning to the end (of the beginning). Free Radic Res 1999;31:261–272.
11 Machlin LJ, Bendich A: Free radical tissue damage: Protective role of antioxidant nutrients. Faseb J 1987;1:441–445.
12 Conn PF, et al: Carotene-oxygen radical interactions. Free Radic Res Commun 1992;16:401–408.
13 Bressler NM, Bressler SB: Preventative ophthalmology. Age-related macular degeneration. Ophthalmology 1995;102:1206–1211.
14 Fukuzawa K, Gebicki JM: Oxidation of α-tocopherol in micelles and liposomes by the hydroxyl, perhydroxyl, and superoxide free radicals. Arch Biochem Biophys 1983;226:242–251.
15 Ozawa T, Hanaki A, Matsuo M: Reactions of superoxide ion with tocopherol and its model compounds: Correlation between the physiological activities of tocopherols and the concentration of chromanoxyl-type radicals. Biochem Int 1983;6:685–692.
16 Sies H, Murphy ME: Role of tocopherols in the protection of biological systems against oxidative damage. J Photochem Photobiol B 1991;8:211–218.
17 Nishikimi M, Machlin LJ: Oxidation of α-tocopherol model compound by superoxide anion. Arch Biochem Biophys 1975;170:684–689.
18 Nishikimi M: Oxidation of ascorbic acid with superoxide anion generated by the xanthine-xanthine oxidase system. Biochem Biophys Res Commun 1975;63:463–468.

19   Sies H: Oxidative stress: From basic research to clinical application. Am J Med 1991;91:31S–38S.

20   Tate DJ, Miceli MV, Newsome DA: Expression of metallothionein isoforms in human chorioretinal complex. Curr Eye Res 2002;24:12–25.

21   Organisciak DT, et al: The protective effect of ascorbic acid in retinal light damage of rats exposed to intermittent light. Invest Ophthalmol Vis Sci 1990;31:1195–1202.

22   Hayes KC: Retinal degeneration in monkeys induced by deficiencies of vitamin E or A. Invest Ophthalmol 1974;13:499–510.

23   Ohta Y, et al: Prolonged marginal ascorbic acid deficiency induces oxidative stress in retina of guinea pigs. Int J Vitam Nutr Res 2002;72:63–70.

24   Marchiafava PL, Longoni B: Melatonin as an antioxidant in retinal photoreceptors. J Pineal Res 1999;26:184–189.

25   Siu AW, Reiter RJ, To CH: The efficacy of vitamin E and melatonin as antioxidants against lipid peroxidation in rat retinal homogenates. J Pineal Res 1998;24:239–244.

26   Siems WG, Sommerburg O, van Kuijk FJ: Lycopene and β-carotene decompose more rapidly than lutein and zeaxanthin upon exposure to various pro-oxidants in vitro. Biofactors 1999;10:105–113.

27   Thomson LR, et al: Long-term dietary supplementation with zeaxanthin reduces photoreceptor death in light-damaged Japanese quail. Exp Eye Res 2002;75:529–542.

28   Thomson LR, et al: Elevated retinal zeaxanthin and prevention of light-induced photoreceptor cell death in quail. Invest Ophthalmol Vis Sci 2002;43:3538–3549.

29   Bernstein PS, Gellermann W: Measurement of carotenoids in the living primate eye using resonance Raman spectroscopy. Methods Mol Biol 2002;196:321–329.

30   Bernstein PS, et al: Resonance Raman measurement of macular carotenoids in normal subjects and in age-related macular degeneration patients. Ophthalmology 2002;109:1780–1787.

31   Touitou Y, et al: Melatonin and aging (in German). Therapie 1998;53:473–478.

32   AREDS Report No 8: A randomized, placebo-controlled, clinical trial of high-dose supplementation with vitamins C and E, β-carotene, and zinc for age-related macular degeneration and vision loss. Arch Ophthalmol 2001;119:1417–1436.

33   Verhoeff FH, Grossman HP: Pathogenesis of disciform degeneration of the macula. Arch Ophthalmol 1937;18:561–585.

34   Friedman E: A hemodynamic model of the pathogenesis of age-related macular degeneration. Am J Ophthalmol 1997;124:677–682.

35   Ciulla TA, et al: Choroidal perfusion perturbations in non-neovascular age-related macular degeneration. Br J Ophthalmol 2002;86:209–213.

36   Mori F, et al: Pulsatile ocular blood flow study: Decreases in exudative age-related macular degeneration. Br J Ophthalmol 2001;85:531–533.

37   Pauleikhoff D, et al: A fluorescein and indocyanine green angiographic study of choriocapillaris in age-related macular disease. Arch Ophthalmol 1999;117:1353–1358.

38   Uretmen O, et al: Color Doppler imaging of choroidal circulation in patients with asymmetric age-related macular degeneration. Ophthalmologica 2003;217:137–142.

39   Vingerling JR, et al: Age-related macular degeneration is associated with atherosclerosis. The Rotterdam Study. Am J Epidemiol 1995;142:404–409.

40   The Eye Disease Case-Control Study Group: Risk factors for neovascular age-related macular degeneration. Arch Ophthalmol 1992;110:1701–1708.

41   Delcourt C, et al: Associations of cardiovascular disease and its risk factors with age-related macular degeneration: The POLA study. Ophthalmic Epidemiol 2001;8:237–249.

42   Hyman L, et al: Hypertension, cardiovascular disease, and age-related macular degeneration. Age-Related Macular Degeneration Risk Factors Study Group. Arch Ophthalmol 2000;118:351–358.

43   Smith W, Mitchell P, Leeder SR: Smoking and age-related maculopathy. The Blue Mountains Eye Study. Arch Ophthalmol 1996;114:1518–1523.

44   McCarty CA, et al: Risk factors for age-related maculopathy: The Visual Impairment Project. Arch Ophthalmol 2001;119:1455–1462.

45   Klein R, et al: The epidemiology of age-related macular degeneration. Am J Ophthalmol 2004;137:486–495.

46   Vingerling JR, et al: Age-related macular degeneration and smoking. The Rotterdam Study. Arch Ophthalmol 1996;114:1193–1196.

47   Seddon JM, Cote J, Rosner B: Progression of age-related macular degeneration: Association with dietary fat, transunsaturated fat, nuts, and fish intake. Arch Ophthalmol 2003;121: 1728–1737.

48   Seddon JM, et al: Progression of age-related macular degeneration: Association with body mass index, waist circumference, and waist-hip ratio. Arch Ophthalmol 2003;121:785–792.

49   Belda Sanchis JI, et al: Are blood lipids a risk factor for age-related macular degeneration? (In Spanish) Arch Soc Esp Oftalmol 2001;76:13–17.

50   Cho E, et al: Prospective study of dietary fat and the risk of age-related macular degeneration. Am J Clin Nutr 2001;73:209–218.

51   Seddon JM, et al: Dietary fat and risk for advanced age-related macular degeneration. Arch Ophthalmol 2001;119:1191–1199.

52   Lotfi K, Grunwald JE: The effect of caffeine on the human macular circulation. Invest Ophthalmol Vis Sci 1991;32:3028–3032.

53   Tomany SC, Klein R, Klein BE: The relation of coffee and caffeine to the 5-year incidence of early age-related maculopathy: The Beaver Dam Eye Study. Am J Ophthalmol 2001;132:271–273.

54   Nygard O, et al: Plasma homocysteine levels and mortality in patients with coronary artery disease. N Engl J Med 1997;337:230–236.

55   Clarke R, et al: Hyperhomocysteinemia: An independent risk factor for vascular disease. N Engl J Med 1991;324:1149–1155.

56   Axer-Siegel R, et al: Association of neovascular age-related macular degeneration and hyperhomocysteinemia. Am J Ophthalmol 2004;137:84–89.

57   Korte GE, Reppucci V, Henkind P: RPE destruction causes choriocapillary atrophy. Invest Ophthalmol Vis Sci 1984;25:1135–1145.

58   Pauleikhoff D, et al: Aging changes in Bruch's membrane. A histochemical and morphologic study. Ophthalmology 1990;7:171–178.

59   Spaide RF, et al: Characterization of peroxidized lipids in Bruch's membrane. Retina 1999;19: 141–147.

60   Green WR, McDonnell PJ, Yeo JH: Pathologic features of senile macular degeneration. Ophthalmology 1985;92:615–627.

61   Ishibashi T, et al: Formation of drusen in the human eye. Am J Ophthalmol 1986;101:342–353.

62   Curcio CA, et al: Accumulation of cholesterol with age in human Bruch's membrane. Invest Ophthalmol Vis Sci 2001;42:265–274.

63   Tokura T, et al: Changes in Bruch's membrane in experimental hypercholesteremia in rats (in Japanese). Nippon Ganka Gakkai Zasshi 1999;103:85–91.

64   Dithmar S, et al: Murine high-fat diet and laser photochemical model of basal deposits in Bruch membrane. Arch Ophthalmol 2001;119:1643–1649.

65   Ishida BY, et al: Regulated expression of apolipoprotein E by human retinal pigment epithelial cells. J Lipid Res 2004;45:263–271.

66   Klaver CC, et al: Genetic association of apolipoprotein E with age-related macular degeneration. Am J Hum Genet 1998;63:200–206.

67   Evans HM, et al: On the existence of a hitherto unrecognized dietary factor essential for reproduction. Science 1922;56:650–651.

68   Stephens RJ, et al: Vitamin E distribution in ocular tissues following long-term dietary depletion and supplementation as determined by microdissection and gas chromatography-mass spectrometry. Exp Eye Res 1988;47:237–245.

69   Wiegand RD, et al: Polyunsaturated fatty acids and vitamin E in rat rod outer segments during light damage. Invest Ophthalmol Vis Sci 1986;27:727–733.

70   Sobrevilla LA, Goodman ML, Kane CA: Demyelinating central nervous system disease, macular atrophy and acanthocytosis (Bassen-Kornzweig syndrome). Am J Med 1964;37:821–828.

71   Muller DP, Lloyd JK, Wolff OH: Vitamin E and neurological function: Abetalipoproteinaemia and other disorders of fat absorption. Ciba Found Symp 1983;101:106–121.

72   Ricciarelli R, et al: α-Tocopherol specifically inactivates cellular protein kinase C α by changing its phosphorylation state. Biochem J 1998;334:243–249.

73   West S, et al: Are antioxidants or supplements protective for age-related macular degeneration? Arch Ophthalmol 1994;112:222–227.

74   Mares-Perlman JA, et al: Serum antioxidants and age-related macular degeneration in a population-based case-control study. Arch Ophthalmol 1995;11:1518–1523.

75   Mares-Perlman JA, et al: Association of zinc and antioxidant nutrients with age-related maculopathy. Arch Ophthalmol 1996;114:991–997.

76   VandenLangenberg GM, et al: Associations between antioxidant and zinc intake and the 5-year incidence of early age-related maculopathy in the Beaver Dam Eye Study. Am J Epidemiol 1998;148:204–214.

77   Roxborough HE, Burton GW, Kelly FJ: Inter- and intra-individual variation in plasma and red blood cell vitamin E after supplementation. Free Radic Res 2000;33:437–445.

78   Taylor HR, et al: Vitamin E supplementation and macular degeneration: Randomised controlled trial. BMJ 2002;325:11.

79   Teikari JM, et al: Six-year supplementation with α-tocopherol and β-carotene and age-related maculopathy. Acta Ophthalmol Scand 1998;76:224–229.

80   Kaiser HJ, et al: Visaline in the treatment of age-related macular degeneration: A pilot study. Ophthalmologica 1995;209:302–305.

81   Li ZY, et al: Amelioration of photic injury in rat retina by ascorbic acid: A histopathologic study. Invest Ophthalmol Vis Sci 1985;26:1589–1598.

82   Eye Disease Case-Control Study Group: Antioxidant status and neovascular age-related macular degeneration. Arch Ophthalmol 1993;111:104–109.

83   Seddon JM, et al: Dietary carotenoids, vitamins A, C, and E, and advanced age-related macular degeneration. Eye Disease Case-Control Study Group. JAMA 1994;272:1413–1420.

84   Mathews-Roth MM: Carotenoids quench evolution of excited species in epidermis exposed to UV-B (290–320 nm) light. Photochem Photobiol 1986;43:91–93.

85   Burton GW, Ingold KU: Beta-carotene: An unusual type of lipid antioxidant. Science 1984; 224:569–573.

86   Bernstein PS, et al: Identification and quantitation of carotenoids and their metabolites in the tissues of the human eye. Exp Eye Res 2001;72:215–223.

87   Sommerburg O, et al: Fruits and vegetables that are sources for lutein and zeaxanthin: The macular pigment in human eyes. Br J Ophthalmol 1998;82:907–910.

88   Albanes D, et al: Alpha-tocopherol and β-carotene supplements and lung cancer incidence in the α-tocopherol, β-carotene cancer prevention study: Effects of baseline characteristics and study compliance. J Natl Cancer Inst 1996;88:1560–1570.

89   The Alpha-Tocopherol, Beta-Carotene Cancer Prevention Study Group: The effect of vitamin E and β-carotene on the incidence of lung cancer and other cancers in male smokers. N Engl J Med 1994;330:1029–1035.

90   Wald G: Human vision and the spectrum. Science 1945;101:653–658.

91   Bone RA, Landrum JT, Tarsis SL: Preliminary identification of the human macular pigment. Vision Res 1985;25:1531–1535.

92   Handelman GJ, et al: Carotenoids in the human macula and whole retina. Invest Ophthalmol Vis Sci 1988;29:850–855.

93   Yemelyanov AY, Katz NB, Bernstein PS: Ligand-binding characterization of xanthophyll carotenoids to solubilized membrane proteins derived from human retina. Exp Eye Res 2001;72:381–392.

94   Landrum JT, Bone RA, Kilburn MD: The macular pigment: A possible role in protection from age-related macular degeneration. Adv Pharmacol 1997;38:537–556.

95   Burton GW: Antioxidant action of carotenoids. J Nutr 1989;119:109–111.

96   Mortensen A, Skibsted LH, Truscott TG: The interaction of dietary carotenoids with radical species. Arch Biochem Biophys 2001;385:13–19.

97   Rapp LM, Maple SS, Choi JH: Lutein and zeaxanthin concentrations in rod outer segment membranes from perifoveal and peripheral human retina. Invest Ophthalmol Vis Sci 2000;41:1200–1209.

98   Dwyer JH, et al: Oxygenated carotenoid lutein and progression of early atherosclerosis: The Los Angeles Atherosclerosis Study. Circulation 2001;103:2922–2927.

99   Mangels AR, et al: Carotenoid content of fruits and vegetables: An evaluation of analytic data. J Am Diet Assoc 1993;93:284–296.

100   Trieschmann M, et al: Macular pigment: Quantitative analysis on autofluorescence images. Graefes Arch Clin Exp Ophthalmol 2003;241:1006–1012.

101   Bone RA, et al: Lutein and zeaxanthin dietary supplements raise macular pigment density and serum concentrations of these carotenoids in humans. J Nutr 2003;133:992–998.

102   Frank RN, Amin RH, Puklin JE: Antioxidant enzymes in the macular retinal pigment epithelium of eyes with neovascular age-related macular degeneration. Am J Ophthalmol 1999;127: 694–709.

103   Karcioglu ZA: Zinc in the eye. Surv Ophthalmol 1982;27:114–122.

104   Tate DJ, et al: Influence of zinc on selected cellular functions of cultured human retinal pigment epithelium. Curr Eye Res 1995;14:897–903.

105   Leure-duPree AE, McClain CJ: The effect of severe zinc deficiency on the morphology of the rat retinal pigment epithelium. Invest Ophthalmol Vis Sci 1982;23:425–434.

106   Vinton NE, et al: Visual function in patients undergoing long-term total parenteral nutrition. Am J Clin Nutr 1990;52:895–902.

107   Stur M, et al: Oral zinc and the second eye in age-related macular degeneration. Invest Ophthalmol Vis Sci 1996;37:1225–1235.

108   Newsome DA, et al: Oral zinc in macular degeneration. Arch Ophthalmol 1988;106:192–198.

109   Ogawa T, et al: Superoxide dismutase in senescence-accelerated mouse retina. Histochem J 2001; 33:43–50.

110   Silverstone BZ, et al: Zinc and copper metabolism in patients with senile macular degeneration. Ann Ophthalmol 1985;17:419–422.

111   Newsome DA, et al: Macular degeneration and elevated serum ceruloplasmin. Invest Ophthalmol Vis Sci 1986;27:1675–1680.

112   Bennett-Richards K, et al: Does oral folic acid lower total homocysteine levels and improve endothelial function in children with chronic renal failure? Circulation 2002;105:1810–1815.

113   Welch GN, Loscalzo J: Homocysteine and atherothrombosis. N Engl J Med 1998;338: 1042–1050.

114   Boushey CJ, et al: A quantitative assessment of plasma homocysteine as a risk factor for vascular disease. Probable benefits of increasing folic acid intakes. JAMA 1995;274:1049–1057.

115   Abu El-Asrar AM, et al: Hyperhomocysteinemia and retinal vascular occlusive disease. Eur J Ophthalmol 2002;12:495–500.

116   Snow KK, Seddon JM: Do age-related macular degeneration and cardiovascular disease share common antecedents? Ophthalmic Epidemiol 1999;6:125–143.

117   Mares-Perlman JA, et al: Dietary fat and age-related maculopathy. Arch Ophthalmol 1995;113: 743–748.

118   Goldberg J, et al: Factors associated with age-related macular degeneration. An analysis of data from the first National Health and Nutrition Examination Survey. Am J Epidemiol 1988;128: 700–710.

119   Klein BE, Klein R: Cataracts and macular degeneration in older Americans. Arch Ophthalmol 1982;100:571–573.

120   Smith W, Mitchell P, Leede SR: Dietary fat and fish intake and age-related maculopathy. Arch Ophthalmol 2000;118:401–404.

121   Mensink RP, Katan MB: Effect of dietary trans fatty acids on high-density and low-density lipoprotein cholesterol levels in healthy subjects. N Engl J Med 1990;323:439–445.

122   Wilson HL, et al: Statin and aspirin therapy are associated with decreased rates of choroidal neovascularization among patients with age-related macular degeneration. Am J Ophthalmol 2004;137:615–624.

123   McGwin G Jr, et al: The association between statin use and age-related maculopathy. Br J Ophthalmol 2003;87:1121–1125.

124   Klein R, et al: Relation of statin use to the 5-year incidence and progression of age-related maculopathy. Arch Ophthalmol 2003;121:1151–1155.

125   Albert CM, et al: Nut consumption and decreased risk of sudden cardiac death in the Physicians' Health Study. Arch Intern Med 2002;162:1382–1387.

126   Kris-Etherton PM, et al: Bioactive compounds in foods: Their role in the prevention of cardiovascular disease and cancer. Am J Med 2002;113(suppl 9B):71–88.

127   Delcourt C, et al: Smoking and age-related macular degeneration. The POLA Study. Pathologies Oculaires Liees a l'Age. Arch Ophthalmol 1998;116:1031–1035.

128   Tamakoshi A, et al: Smoking and neovascular form of age-related macular degeneration in late middle-aged males: Findings from a case-control study in Japan. Research Committee on Chorioretinal Degenerations. Br J Ophthalmol 1997;81;901–904.

129   Christen WG, et al: A prospective study of cigarette smoking and risk of age-related macular degeneration in men. JAMA 1996;276:1147–1151.

130   Seddon JM, et al: A prospective study of cigarette smoking and age-related macular degeneration in women. JAMA 1996;276:1141–1146.

131   Goldfinger TM: Beyond the French paradox: The impact of moderate beverage alcohol and wine consumption in the prevention of cardiovascular disease. Cardiol Clin 2003;21:449–457.

132   Renaud SC, et al: Wine, beer, and mortality in middle-aged men from eastern France. Arch Intern Med 1999;159:1865–1870.

133   Obisesan TO, et al: Moderate wine consumption is associated with decreased odds of developing age-related macular degeneration in NHANES-1. J Am Geriatr Soc 1998;46:1–7.

134   Ajani UA, et al: A prospective study of alcohol consumption and the risk of age-related macular degeneration. Ann Epidemiol 1999;9:172–177.

135   Cho E, et al: Prospective study of alcohol consumption and the risk of age-related macular degeneration. Arch Ophthalmol 2000;118:681–688.

136   De Lorimier AA: Alcohol, wine, and health. Am J Surg 2000;180:357–361.

137   Klein R, et al: The relationship of age-related maculopathy, cataract, and glaucoma to visual acuity. Invest Ophthalmol Vis Sci 1995;36:182–191.

138   Attebo K, Mitchell P, Smith W: Visual acuity and the causes of visual loss in Australia. The Blue Mountains Eye Study. Ophthalmology 1996;103:357–364.

139   Klaver CC, et al: Age-specific prevalence and causes of blindness and visual impairment in an older population: The Rotterdam Study. Arch Ophthalmol 1998;116:653–658.

140   The Diabetes Control and Complications Trial Research Group: The effect of intensive treatment of diabetes on the development and progression of long-term complications in insulin-dependent diabetes mellitus. N Engl J Med 1993;329:977–986.

141   UK Prospective Diabetes Study (UKPDS) Group: Intensive blood-glucose control with sulphonyl-ureas or insulin compared with conventional treatment and risk of complications in patients with type 2 diabetes (UKPDS 33). Lancet 1998;352:837–853.

142   Klein BE, et al: The Wisconsin Epidemiologic Study of Diabetic Retinopathy. XIII. Relationship of serum cholesterol to retinopathy and hard exudate. Ophthalmology 1991;98: 1261–1265.

143   Chew EY, et al: Association of elevated serum lipid levels with retinal hard exudate in diabetic retinopathy. Early Treatment Diabetic Retinopathy Study (ETDRS) Report 22. Arch Ophthalmol 1996;114:1079–1084.

144   Lyons TJ, et al: Diabetic retinopathy and serum lipoprotein subclasses in the DCCT/EDIC cohort. Invest Ophthalmol Vis Sci 2004;45:910–918.

145   Sen K, et al: Simvastatin retards progression of retinopathy in diabetic patients with hypercholes-terolemia. Diabetes Res Clin Pract 2002;56:1–11.

146   Lu M, et al: Advanced glycation end products increase retinal vascular endothelial growth factor expression. J Clin Invest 1998;101:1219–1224.

147   Kim I, et al: Vascular endothelial growth factor expression of intercellular adhesion molecule 1, vascular cell adhesion molecule 1, and E-selectin through nuclear factor-κB activation in endothe-lial cells. J Biol Chem 2001;276:7614–7620.

148   Mamputu JC, Renier G: Advanced glycation end-products increase monocyte adhesion to retinal endothelial cells through vascular endothelial growth factor-induced ICAM-1 expression: Inhibitory effect of antioxidants. J Leukoc Biol 2004;75:1062–1069.

149   Obrosova IG, et al: Antioxidants attenuate early upregulation of retinal vascular endothelial growth factor in streptozotocin-diabetic rats. Diabetologia 2001;44:1102–1110.

150   Baynes JW: Role of oxidative stress in development of complications in diabetes. Diabetes 1991;40:405–412.

151   Augustin AJ, et al: Correlation of blood-glucose control with oxidative metabolites in plasma and vitreous body of diabetic patients. Eur J Ophthalmol 2002;12:94–101.

152 Kowluru RA, Tang J, Kern TS: Abnormalities of retinal metabolism in diabetes and experimental galactosemia. VII. Effect of long-term administration of antioxidants on the development of retinopathy. Diabetes 2001;50:1938–1942.

153 Di Leo MA, et al: Potential therapeutic effect of antioxidants in experimental diabetic retina: A comparison between chronic taurine and vitamin E plus selenium supplementations. Free Radic Res 2003;37:323–330.

154 Millen AE, et al: Relation between intake of vitamins C and E and risk of diabetic retinopathy in the Atherosclerosis Risk in Communities Study. Am J Clin Nutr 2004;79:865–873.

155 Millen AE, et al: Relations of serum ascorbic acid and α-tocopherol to diabetic retinopathy in the Third National Health and Nutrition Examination Survey. Am J Epidemiol 2003;158:225–233.

156 Mayer-Davis EJ, et al: Antioxidant nutrient intake and diabetic retinopathy: The San Luis Valley Diabetes Study. Ophthalmology 1998;105:2264–2270.

157 Bursell SE, et al: High-dose vitamin E supplementation normalizes retinal blood flow and creatinine clearance in patients with type 1 diabetes. Diabetes Care 1999;22:1245–1251.

Ursula Schmidt-Erfurth
Universitätsklinik der Augenheilkunde und Optometrie
Währinger Gürtel 18–20, AT–1090 Vienna (Austria)
Tel. +43 140 4007930, Fax +43 140 4007932,
E-Mail ursula.schmidt-erfurth@akhwien.at

Author Index

·····················
Subject Index

Vitamin E (continued)
  disease prevention studies
    age-related macular degeneration 64,
      126–128
    cancer 10
    cardiovascular disease 9, 10
    cataract 10, 63, 109, 113
    infection and immune response 10
  food sources 7, 8
  functions
    α-tocopherol 7, 8
    γ-tocopherol 8
  metabolism 63
  recommended dietary allowance 9
  safety of supplementation 10, 11
  synergism with other antioxidants
    65, 66
  tissue levels 63

Zeaxanthin, *see* Carotenoids
Zinc
  deficiency 25, 26
  disease prevention studies
    age-related macular degeneration 28,
      29, 132
    growth and development 27
    infection and immune response 27, 28
    pregnancy complications 28
  food sources 24
  functions 24, 25
  interactions with other nutrients
    calcium 25
    copper 13, 25
    folic acid 25
    iron 25
  recommended dietary allowance 26, 27
  safety of supplementation 29